TRANSACTIONS

OF THE

AMERICAN PHILOSOPHICAL SOCIETY

HELD AT PHILADELPHIA
FOR PROMOTING USEFUL KNOWLEDGE

———

NEW SERIES—VOLUME 39, PART 1
1949

———

THE WHOLE LIFE CYCLE OF CHROMOSOMES AND THEIR COILING SYSTEMS

L. R. CLEVELAND

Professor of Biology
Harvard University

———

THE AMERICAN PHILOSOPHICAL SOCIETY
INDEPENDENCE SQUARE
PHILADELPHIA 6

JUNE, 1949

THE WHOLE LIFE CYCLE OF CHROMOSOMES
AND THEIR COILING SYSTEMS

L. R. CLEVELAND

CONTENTS

INTRODUCTION

Chromosomes play so important a role in heredity and development that it is desirable to know their entire life cycle since this knowledge will provide a basis for a better understanding of their behavior. Heretofore, owing more perhaps to a lack of suitable material than anything else, only a portion of their life cycle has been studied. That place in the cycle where duplication occurs has really not been studied, notwithstanding the fact that many accounts of the division of chromosomes have been published. These accounts, for the most part, deal with late stages in the separation of daughter chromosomes, not the duplication of the parent chromosome.

The idea that chromosomes disappear and reappear in each cell generation is still fairly common, even among biologists, although genetics showed long ago that chromosomes must be continuous from one generation to another. The proof of this statement, if anyone doubts it, may be seen in the following quotations from two papers, the like of which are still being published in fairly great numbers: Rafalko (1947) "The chromosomes disintegrate into tiny particles"; Noble (1947) "The Feulgen-positive reaction of the endosome, and its tendency to break up and mingle with periendosomal granules suggests that the endosome takes part in chromosome formation."

During the last twenty years several hundred papers have appeared describing minor, major, and (in a few instances) super coils of chromosomes. But these coiling systems, like the chromosomes, have been studied in only a small portion of their life cycle. What happens to them during most of the cycle, how and when they are produced, how and when they disappear, and many other events in their life cycle are unknown.

It is the object of this paper not only to clear up these unknown events in the life cycle of chromosomes but also to point out certain erroneous conclusions that have been made regarding the manner in which relational coiling is produced and disappears, the relation of centromere to chromosome, relic coiling, matrix, and chromonema.

It is also the aim of this paper to give such an accurate and complete account of all the events in the life cycle of chromosomes and their coiling systems that it will dispel some of the vague generalities and sheer speculations that have been tossed about for many years without really having advanced our knowledge of cytology very much if at all.

The results reported here are based on detailed studies of the chromosomes of seven families and forty genera of flagellates. However, only a very small amount of the material studied is considered in this paper. The studies were completed several years ago but were not published because before doing so I wanted to compare these chromosomes with those of other classes of protozoa and of higher animals and plants to determine whether all chromosomes are fundamentally alike or whether these chromosomes of flagellates belong in a special class to themselves. This comparison—together with the initial task—was so long and arduous that several of my colleagues thought it would never be completed. However, it was worth while since I am now able to say with definite assurance that all chromosomes belong to either one or the other of the two types described here (single or double chromatids).

Major coils can be seen in many types of cells in animals and plants when the material has been properly fixed and stained or when viewed in the living condition with good phase contrast equipment; and minor coils may also be seen clearly, in a fair number of cells, under the same conditions, by the eye accustomed to looking for them, but the whole life cycle of chromosomes can be observed in very few cells. The number of chromosomes must be small and the task is easier when they are rod-shaped. If every detail is to be seen with unmistakable clarity in all phases of the life cycle, the number of chromosomes cannot be more than two, and even then it is a great help, in those stages when only minor coils are present, if there is some simple, clear-cut method of distinguishing one chromosome from the other. Certain species of the genera *Holomastigotoides* and *Spirotrichonympha* fulfill all of these exacting conditions perfectly. In addition, one large species of *Holomastigotoides*, *H. tusitala* species novum, has an achromatic figure, centromeres, and centrioles which can be seen clearly throughout the life cycle of the chromosomes. This, together with the fact that in this (and in some re-

1

lated) species the chromosomes are always anchored to the centrioles, that each of the two chromosomes has a terminal nucleolus, and one, the longer chromosome, has a lateral nucleolus, makes it possible to identify any portion of either chromosome in any stage of its life cycle. For these reasons this organism has been used mostly in illustrating the life cycle of chromosomes (a few drawings of other species and genera have been included to show the uniformity of chromosomal morphology). Of the slightly over 500 species of flagellates studied, a few were almost as good as *H. tusitala;* but most of them were not. However, in all of them—even those with a large number of chromosomes—it is perfectly clear that the chomosomes go through precisely the same stages of development as those of this organism, despite the fact that each chromosome cannot be traced individually from telophase to telophase. The same, of course, is also true with cells of higher forms.

Chromosomes vary in shape from those of two arms of equal length to those of a single arm, the shape depending on the position of the centromere. Despite the fact that the existence of terminal centromeres has been questioned by many cytologists, and in some instances categorically denied, they very definitely occur.

Since there is no general agreement among cytologists as to when a daughter chromosome or chromatid becomes a chromosome, it is desirable for me to state how I am using these terms, and to use them uniformly throughout the paper (unfortunately some authors do not). As soon as duplication in a chromosome or a chromatid occurs, each part is as much of a chromosome as it will ever be, and should be so termed; but long-established usage makes it difficult to employ this terminology. The next point, when the daughters might be termed chromosomes, is after they have lost their relational coiling, but are still in the parent nucleus. After this, another place where a good many cytologists begin to call them chromosomes, is when nuclear division is completed; but if this is done before reorganization of the nuclear membrane is accomplished, one might as well call them chromosomes as soon as they have separated. On the other hand, if one waits until a new nuclear membrane is organized, in some instances one will not call them chromosomes until cell division is accomplished. This point is objectionable then because the terminology would not apply uniformly to all cells. Perhaps the least objectionable point at which to begin calling them chromosomes is after cell division, but there are difficulties even here when one is dealing with multinucleate cells or syncitia. Such terms as daughter chromosomes, chromatids, and half chromatids would be unnecessary—in fact need never have been coined— if we would only admit that a cell does not have the same number of chromosomes all the time, that after it divides it has half the number that it has later after the chromosomes are duplicated. In the terminology I am

employing a chromatid does not become a chromosome until after cell division.

It has also been difficult for me to choose between the usage of daughter chromosome and chromatid. It certainly seems desirable in meiosis, where chromosomes from two parents are concerned, to use chromatid; and in mitosis, either haploid or diploid, to use daughter chromosome; but the problem is not as simple as this. For example, when a haploid cell becomes diploid due to endomitosis and then by meiosis returns to a haploid, the chromosome set-up is the same as in mitosis, yet tetrads are formed and crossing over occurs just as in either gametic or zygotic meiosis. And, in addition to this, we are finding more and more examples all the time of diploid mitosis where tetrads and crossing over may be seen plainly. In view of these facts, and others which will appear later, it seems desirable to me to use chromatid both in meiosis and mitosis. If daughter chromosome is used regard it as a synonym of chromatid.

Practically all the well-known fixatives and stains have been used. Schaudinn's and Flemming's gave slightly better results than other fixatives. Heidenhain's iron alum hematoxylin,[1] Feulgen, crystal violet, and Bodian's stains were used most. Either Feulgen or crystal violet stains the chromosomes of all cells in a preparation more uniformly, but hematoxylin, when properly prepared and handled, has a far greater degree of differentiation and therefore will show minor coils more plainly when these are incorporated in close-together, large, major coils of metaphase chromosomes. The other stains, in my hands, tend to fill in the spaces between the minors in the tight majors and thus show the majors more plainly than hematoxylin, but the minors less plainly. Bodian's, except in a few organisms and following certain fixatives, stains chromosomes poorly, but no other stain stains most extranuclear organelles plainly at all in comparison to it.

I cannot express the depth of my indebtedness to Mrs. Laura O. Samuelson for the care and time—better part of six years—she has taken in making the drawings.

I am also greatly indebted to Dr. Harold Kirby for preparations of the protozoa of several species of termites which, together with my own preparations, enable me to make certain generalizations regarding distribution and types of Spirotrichonymphidae which I otherwise could not make.

Dr. E. M. Miller, of the University of Miami, rendered valuable assistance by keeping me well supplied with living *Prorhinotermes simplex* Hagen.

------

[1] I keep a 10 per cent stock solution in 95 per cent ethyl alcohol. It is diluted with distilled water to 1 per cent before it is used and no time is allowed for ripening. The preparations are transferred directly from the 4 per cent iron alum to the stain without washing off any of the adhering iron alum. In this way the stain ripens as it stains and should be discarded within a day or two. The staining time is 30 to 60 minutes and destaining is carried out rather quickly in 4 per cent iron alum.

This investigation has been supported by grants from the Penrose Fund of the American Philospohical Society and the Milton Fund of Harvard University.

THE GENUS *HOLOMASTIGOTOIDES*

This genus was established by Grassi and Foa in 1911 at which time they described a single species, *H. hertwigi* from *Coptotermes hartmanni* Holmgren from Brazil. Grassi (1917) described two species, *H. mirabilis* from *Coptotermes sjöstedt* Holmgren from French Guiana and *H. hemigymnum* from *Coptotermes lacteus* Froggatt from Australia. Koidzumi (1921) described *H. hartmanni* from *Coptotermes formosanus* Shiraki from Formosa. MacKinnon (1926) studied from *Heterotermes* (= *Leucotermes*) *tenuis* Hagen from Trinidad what she thought desirable (in view of Grassi's single figure and very brief description of *H. hemigymnum*) to regard temporarily as *H. hemigymnum*. I have preparations of the protozoa of both termites and the organism MacKinnon studied is clearly a new species.

De Mello (1942) says he has found 13 species of *Holomastigotoides* in India in *Heterotermes* (= *Leucotermes*) *indicola* and in unidentified species of *Heterotermes* and *Coptotermes*, most of the species being, in his opinion, new. Several species of *Holomastigotoides* have been described from *Hodotermes*, but these, in my opinion, belong in another genus. No one has described any protozoa from *Prorhinotermes* or *Psammotermes*, and there are many species of *Holomastigotoides* in each genus. There are undescribed organisms in *Schedorhinotermes* which, like those of *Hodotermes*, are fairly closely related to *Holomastigotoides*, but which, in my opinion, should be placed in another (new) genus. In *Rhinotermes* there are no forms related to *Holomastigotoides*. There are also no organisms known from *Reticulitermes* which belong in this genus, but a related genus is present. This, then, so far as our present information goes, restricts *Holomastigotoides* to four genera of termites: *Prorhinotermes*, *Heterotermes*, *Coptotermes* (figs. 46, 48, 51)[2], and *Psammotermes* (fig. 47). I have studied the protozoa of many species of each of these genera and in each species there are always several species of *Holomastigotoides*. The question of species in this genus presents many unusually interesting problems which I hope sometime to find time to consider.

Holomastigotoides, like most genera of the family Spirotrichonymphidae, has bands from which the flagella arise, which spiral a good portion of the body anterioposteriorly, and which are known as flagellar bands. The degree of spiralization depends on the size of the cell and the number of bands present. There are small species with only two bands, and large ones with as many as sixty. There are many species with numbers between two and sixty. When many bands are present,

as in several of the large species from *Psammotermes*, *Coptotermes*, and *Heterotermes*, there is not room for them to spiral the body more than once, or at most twice; but, when fewer bands are present, they may spiral the body several times, each spiral of the body being equivalent to one gyre of coiling in a chromosome. These spiral bands, like chromosomes in certain stages in their life cycle, are also relationally coiled and have to unwind in order to separate. This presents special problems for a cell when it undergoes cytoplasmic division. A few genera of the Spirotrichonymphidae have solved the problem by discarding the old bands and developing new ones during each cell generation, but most genera, including *Holomastigotoides*, retain the bands of the parent cell and pass them on to daughters, each daughter sometimes receiving the same number of bands, sometimes not. And in *Spirotrichonympha bispira* (Cleveland, 1938) the parent cell retains both bands and no unwinding occurs. One centriole migrates to the posterior end of the cell where it produces new bands for the small, posterior daughter. However, in the closely related species, *S.-polygyra*, the bands, of which there are four, unwind and two go with each daughter. A study of the various types of behavior of these flagellar bands in several hundred species of flagellates has helped me greatly in understanding chromosomes. They have individual and relational coiling just as chromosomes do, and they lose relational coiling in the same manner as chromosomes, by unwinding; but the relational coiling of the bands is not produced by the same type of duplication as occurs in chromosomes; a band (or bands) grows out from a centriole and follows an old band, thus acquiring a spiral like the old one as it increases in length. But a new band will also spiral as it extends across the body when no old band is present. Its individual spirals or coils, like the minor coils of chromosomes, are due to its inherent molecular structure, and when this substance is duplicated, irrespective of whether duplication is linear, as in the spiral bands, or lateral, as in chromosomes, the new band or the new chromosome spirals individually like its mate or parent, and the two, bands or chromosomes, are therefore relationally coiled. Relational coiling in chromatids, then, means that each of the two relationally coiled chromatids have identical individual minor coiling systems, that the parent chromosome which produced them had the same system and, in duplicating itself, passed its minor coiling system unchanged to its daughters. The existence of relational coiling, either of chromatids or half chromatids, means minor coils were present when duplication occurred. In fact, it is not necessary to be able to trace the minor coils from one generation to another—or even to see them at all—to prove that they are present at duplication, because the presence of relational coiling is sufficient proof in itself of the existence of an individual coiling system. Then, since majors and supers disappear some time before duplication occurs, the individual coils that are

---

[2] All figures are in the plates at the end of the paper; text-figures appear in the body of the text.

present and are responsible for the relational coiling must be minors.

In all species of *Prorhinotermes,* all species of *Holomastigotoides* have rod-shaped chromosomes; the same is true for all species of *Heterotermes* except one, *H. platycephalus* from S.W. Australia. Here the situation is the same as in many species of *Coptotermes,* namely, one rod-shaped and one V-shaped chromosome in two-chromosome species. In all species of *Psammotermes,* the two-chromosome organisms have a V and a J. In all four of these genera of termites the basic form of *Holomastigotoides* has two chromosomes; however aneuploidy and polyploidy have occurred throughout the group, producing forms with three and four chromosomes in all four genera and, in *Psammotermes,* forms with five chromosomes. It is perhaps undesirable at present on the basis of the chromosomes alone to regard these forms with more than two chromosomes as species, although they are larger, particularly those with four and five chromosomes.

There are also two types of cytoplasmic division, transverse and longitudinal, which occur in all the forms with two and three chromosomes, but only longitudinal division occurs in the four and five chromosome forms. Evidently the longitudinal type of division, which produces daughter cells of equal size and to which an equal number of parent flagellar bands are passed on, is the original and more general type of division; and the transverse type, which produces daughter cells of unequal size except in forms with only two bands, should be regarded as of more recent origin, even though it arose prior to the Tertiary (because it occurs in *Holomastigotoides* from all species of these four genera of termites which have been separate and distinct for a longer period). In the transverse type of division, one flagellar band becomes free of the other bands (or band) anteriorly and as it unwinds it moves slowly posteriorly, finally becomes completely separated from the other bands and occupies the small, posterior portion of the cell which, prior to the beginning of division, was free of flagellar bands. Here this band tightens its coils and the cytoplasm slowly constricts between it and the remaining anterior bands; and finally a small posterior daughter cell with a single flagellar band becomes free (figs. 34–39, 78–80). Then begins a fairly long period of growth of cytoplasm and formation of new bands so that eventually this cell is as large as the parent cell from which it separated and has the same number of flagellar bands. Meanwhile the anterior daughter develops a single new band so that it too has the same number as the parent cell.

Although there is considerable difference in individual termites in the number of transverse dividing and longitudinal dividing organisms present, on the whole the two types occur in approximately equal numbers. Even though there is a marked and constant difference in these forms in the position of the lateral nucleolus

and the size and number of major coils, I have tried, at times, to conceive of some plan whereby one form might produce the other; such, for example, as a five-banded transverse dividing form and an eight-banded longitudinal dividing form. If a five-banded form produces an anterior daughter with four bands and a posterior daughter with one band, then, if the four-banded form produced four new bands, this would give a cell with eight bands, the number in the longitudinal dividing species; and, on the other hand, if the one-banded form produced four new bands, this would make five, the number in the transverse dividing species. But this sort of scheme is clearly unworkable since it would soon exhaust the supply of transverse dividing forms. Also the fact that diploids divide only longitudinally suggests that each type is independent.

In *Holomastigotoides* centrioles always follow two adjacent flagellar bands irrespective of the number of bands present. In transverse dividing forms, they follow the bands to a point near the nucleus; here, in many species, they leave the bands and, by means of an ever present—either in whole or in part—achromatic figure, connect with the nucleus and thus serve to keep it always in precisely the same position in a cell. In longitudinal dividing forms, the centrioles are very short and extend along the flagellar bands only a short way from the anterior end toward the nucleus (figs. 40, 90). In these forms the nucleus, except for axostyles that surround it, is free in the cytoplasm; and unlike that of the transverse dividing forms it has no connection with the early achromatic figure, which, even though always present, lies a considerable distance from the nucleus in the prophase, much too far to make connection with the chromosomes (figs. 40, 44, 45, 84, 85).

Only two workers, Koidzumi (1921) and MacKinnon (1926), have written anything regarding mitosis in *Holomastigotoides,* and neither saw any stages in division. What they thought were dividing forms, even though all alike, were actually resting prophases. What they did not know, nor I either for some time, was that the chromosomes of *Holomastigotoides* and several other genera of flagellates rest in the prophase. The same is true for other groups of protozoa, and is not unknown in higher forms. In some genera resting nuclei are in the telophase, but in most cells, protozoa as well as higher forms, the resting stage is when chromosomes are in the apparently longest stage of their life cycle, that is, when only minor coils are present. This stage is generally known as "interphase," although if "interphase" and "resting" are regarded as synonyms—they should not be—the same term is really just as applicable to those organisms where resting nuclei have prophase or telophase chromosomes. Much confusion and misunderstanding will be avoided if resting stage is used for the stage when chromosomes are undergoing little if any activity, irrespective of whether they are in the prophase, telophase, or interphase stage of their life cycle.

SPECIES OF *HOLOMASTIGOTOIDES* IN *PRORHINOTERMES*

The species of *Holomastigotoides* are the same in all the species of *Prorhinotermes* that I have examined, and this includes species from Florida, Madagascar, and the Dutch East Indies, which species have been separated for a considerable period of time. Hence, the many interesting features of these flagellates are, to say the least, not of recent origin.

The two largest species, *H. tusitala* sp. nov., which divides transversely, and *H. diversa* sp. nov., which divides longitudinally, are described briefly in later sections of this paper, and need not be considered here. All the other species are smaller, have smaller chromosomes, and with one exception, have fewer flagellar bands; and, since there is no difficulty in differentiating them from the two larger species whose chromosomes receive most consideration, detailed descriptions, necessary for naming them, seem unwarranted at this time.

If these smaller organisms were divided into species, it would be done mostly on the number of flagellar bands and the type of cytoplasmic division, longitudinal or transverse. Of those that divide transversely, there are three species with less than five flagellar bands, the number in *H. tusitala;* and of those that divide longitudinally, there are at least three, possibly four, species with fewer than eight flagellar bands, the number in *H. diversa*. The same types of haploids, aneuploids, and diploids occur in these species as in *H. tusitala* and *H. diversa* where they are described. All species, irrespective of the number of chromosomes or the type of division, exist in two types, namely those with single and those with double chromatids (compare figs. 44 with 84 and 45 with 85). Single and double chromatids also occur in *Holomastigotoides* from *Coptotermes, Heterotermes,* and *Psammotermes;* in *Spirotrichonympha* (figs. 49, 50, 86), and in several other genera. These types have either arisen independently in each genus or they are older than these genera of termites—very probably the latter. Although there is considerable difference in the size of the major coils of the single and double chromatids, and certainly no possibility of a chromatid changing from one type to the other quickly or easily, it seems somewhat undesirable to me to make species on this basis alone, and there are clearly no other observable differences. Perhaps this is an example of evolution in chromosomes which has not produced noticeable cytoplasmic changes. However, the small posterior daughter cells —produced by unequal transverse division of the cytoplasm—of one type never attempt to enter dividing cells of the other type, nor do posterior daughters, of one type, ever attempt to fuse with those of the other type, (fig. 91). Hence, cells with single chromatids must be quite distinct cytoplasmically from those with double chromatids.

These posterior daughters, which partially fuse with themselves, with dividing cells in the process of producing posterior daughters, and occasionally with anterior daughters, may represent the functionless remains of a once well established sexual cycle, which has become greatly distorted and mostly suppressed through the long period of evolutionary changes, morphological and physiological, in their termite hosts since they diverged from their problattoid ancestors (figs. 91, 93). They may represent gametes which are no longer able to fuse completely and hence to carry out their former function of fertilization, as their close relatives, *Leptospironympha* and *Macrospironympha,* are still able to do in their blattid host, *Cryptocercus,* in which far fewer morphological and physiological changes have occurred than in termites (Cleveland, 1947a, 1947b). If these assumptions are correct, these cells, which once functioned as gametes can only survive now by developing new flagellar bands, the same number as the parent cell which produced them—followed by a long period of cytoplasmic growth. Observations indicate that most of them reproduce asexually in this manner; a few may fail to survive. However, if most of them did not survive, the longitudinal dividing species, which produce nothing even remotely resembling gametes, either in structure or function, would soon overgrow the transverse dividing species if only half of their progeny survived.

In the protozoa of termites, as in those of *Cryptocercus,* most cell division is correlated with molting in the host. In *Cryptocercus,* however, the protozoa produce either gametes, gametic nuclei, or gametic chromosomes under the influence of the roach molting hormone, some genera reacting in one way, some in another to the hormone. In termites, so far as we know, all protozoa, except a few small species of polymastigote flagellates, are lost each time the insect molts, and have to be regained from the non-molting members of the colony. This is true, according to Dr. Max Day who recently made observations for me, even in the most primitive, roach-like termite, *Mastotermes darwiniensis*. However, since the hypermastigote flagellates of roaches and termites are undoubtedly of common origin and are still very closely related, it would be surprising not to find somewhere, in the vast array of species and genera of termites that exist today, at least a remnant in some form of the sexuality which existed in this group of flagellates for millions of years prior to the divergence of termites and roaches. Perhaps this is what I am seeing in *Holomastigotoides*. Only more study, particularly shortly before and after molting, will give a definite explanation to the unusual behavior observed in the transverse dividing species.

HOLOMASTIGOTOIDES TUSITALA SPECIES NOVUM

Since the organism whose chromosomes are given most consideration in this paper is new and possesses several hitherto undescribed and uniquely interesting features, considerable time, I feel, should be devoted to it before taking up a detailed description of the life cycle of its chromosomes. If the extranuclear organelles of the cell are understood fully, then it will be easier to follow the description of the chromosomes.

This is the largest of the transverse dividing species in *Prorhinotermes*. It exists in forms with two and in forms with three chromosomes. All chromosomes have terminal centromeres, and, of course, are rod-shaped. In the form with two, one is long and one is short; both have terminal nucleoli; and the long one has a lateral nucleolus, which is located about one-third of the distance from the centromere toward the terminal nucleolus, considerably farther from the centromere than in the largest longitudinal dividing species, *H. diversa*. The chromosomes at late prophase and metaphase are longer and narrower—because they have more and smaller major coils—than those of *H. diversa* (compare figs. 40–

45 and 20*d*–25*d*). The forms with three chromosomes may be composed of two short ones and one long one or of two long ones and one short one (figs. 81, 82). No transverse dividing forms have become permanent diploids. A few have been seen, but they are so scarce that they perhaps should be regarded as anomalies with a temporary existence and possibly produced by nondisjunction. Then, the two-chromosome form and each combination of the three-chromosome aneuploids occur in single and double chromatid types, making in all six types or races in this species.

All the surface of the cell is covered with flagella (text-fig. 1, *a*) which are directed posteriorly through the cytoplasm (text-fig. 1, *d*). Those arising from near the posterior end of the body extend beyond it. The flagella arise from five flagellar bands, each of which spirals the body 5½ times (text-fig. 1, *b*, *e*). That is, each band

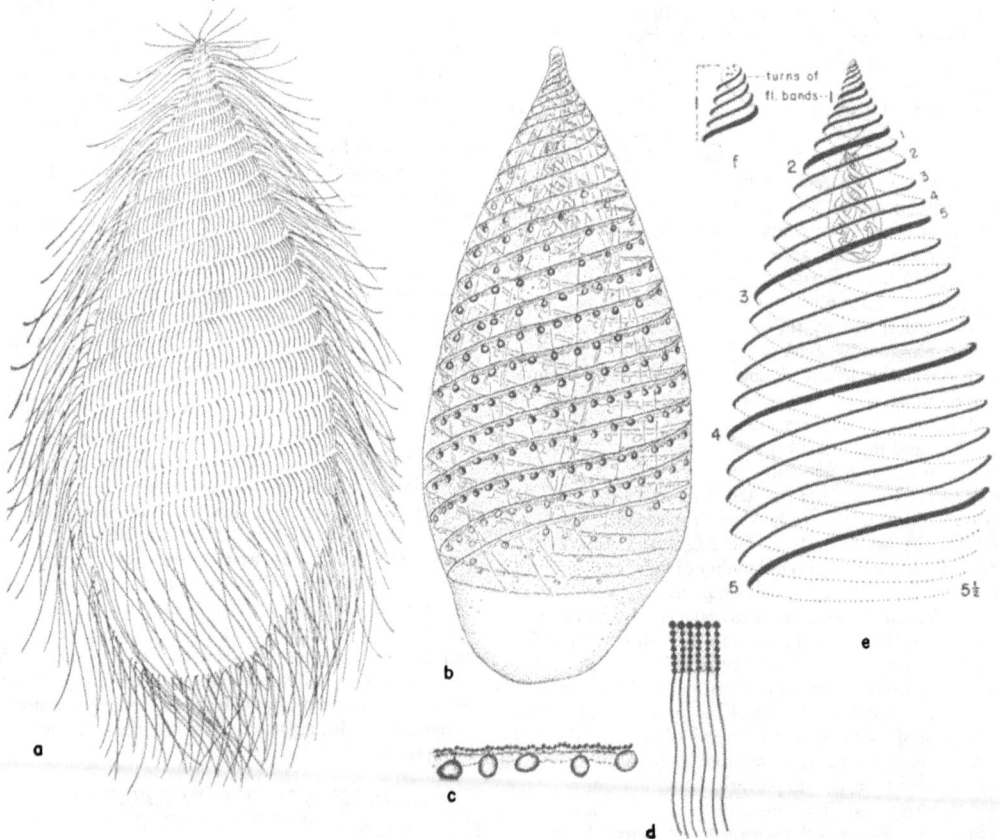

TEXT-FIG. 1. *Holomastigotoides tusitala.* (*a*) Surface view of flagella, those at sides and posterior end being shown full length. (*b*) Flagellar bands, parabasal bodies along bands, long, thin axostyles extending anterio-posteriorly for most of body length. (*c*) Detail of parabasals, flagella-bearing and supporting portions of band. (*d*) Detail of flagella as they leave band. (*e*) Flagellar bands, nucleus, centrioles following bands for 1½ turns. Band 5 drawn heavy to show number of turns bands make around body. (*f*) Detail of anterior end showing one turn of each band.

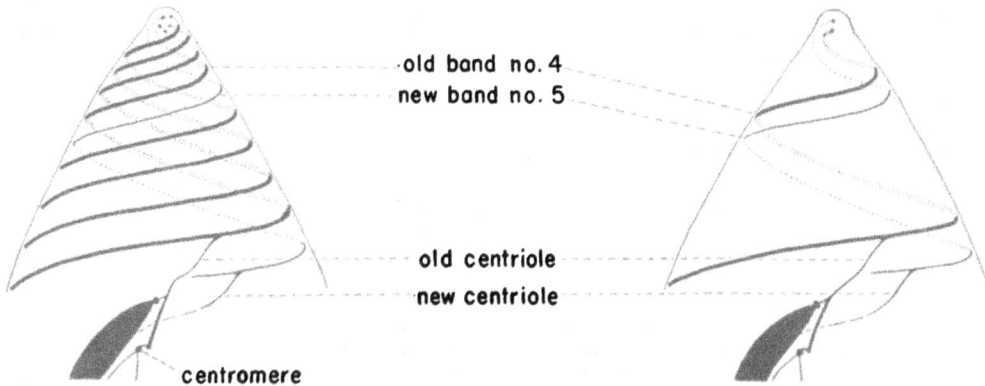

old band no. 4
new band no. 5
old centriole
new centriole
centromere

TEXT-FIG. 2. *Holomastigotoides tusitala.* Interrelationship of flagellar bands, centrioles, centromeres, and chromosomes.

has 5½ gyres or turns of individual coiling. The first turn is smaller than the second and the second smaller than the third. The third, fourth, and fifth turns are about the same size (text-fig. 1, *e*). Observations on living cells show that these bands are quite rigid and hence are responsible for the characteristic bullet-shaped anterior end of the cell. They all arise from near the same point in the anterior end (text-fig. 1, *f*).

Each band is composed of two portions: a fairly narrow, more heavily staining, anterior portion from which the flagella arise, and which is termed the flagella-bearing portion; and a lighter, broader, underneath and posterior portion, which is termed the supporting portion (text-fig. 1, *b, c*). In living material the flagella-bearing portion is denser and more conspicuous than the supporting portion. In fact, the supporting portion may be overlooked unless proper light filters are used.

Slightly over and posterior to the supporting portion lie the parabasal bodies. They occur at very regular intervals along each flagellar band, beginning with the third turn or thereabouts and continuing almost to the posterior ends of the bands (text-fig. 1, *b*). They arise from a thread, the parabasal thread, which lies posterior to the flagella-bearing portion and which adheres to the upper surface of the supporting portion. Each body is connected to the parabasal thread by a stem or thread of its own (text-fig. 1, *c*).

Fine, fibrillar axostyles arise from underneath the flagellar bands near the anterior end and extend, between the bands, posteriorly almost to the end of the body (text-fig. 1, *b*).

Centrioles arise at the anterior end of the body and follow closely along the flagella-bearing portion of bands 4 and 5 for 1½ turns. Then they leave the bands and extend inwardly for a short distance to the anterior margin of the nucleus, where they produce the ever present achromatic figure, which is connected to the centromeres of the chromosomes from one generation to another (text-fig. 1, *e;* text-fig. 2). Thus the chromosomes are always connected with the flagellar bands, a very necessary feature when one considers the unusual and interesting manner of cell division in this and related species of *Holomastigotoides.* Soon after nuclear division is completed, flagellar band 5 becomes free of the other four bands in the anterior end and slowly begins to unwind (fig. 34*A*). In order to unwind completely and thus free itself entirely of the other bands, this band must turn around or rotate the body of the cell 5½ times (figs. 34–39). This band also must carry a nucleus with it; otherwise the cell of which it eventually becomes a part would have no nucleus, unless, of course, the nucleus of its own accord migrated to the posterior end of the body at just the right time, a difficult undertaking, which, I think, would misfire most of the time. It is really impossible, of course, to be certain which does the migrating, the nucleus or the flagellar band, since the two always are connected by the centriole and achromatic figure. I suppose the nucleus could do the moving and the band would have to follow it; but, if this is true, why does the nucleus rotate so as to unwind the band? If a centriole for most of its length did not adhere closely to this band, the connection between the band and nucleus could not be maintained, and most attempts at cell duplication would result in failure. The fibres of the achromatic figure must persist, otherwise the centriole could not maintain its connection with the centromeres of the chromosomes. But it is all these unusual features and requirements of this organism that make it so valuable in studying chromosomes and structures related to them. Not only do centromeres have to exist from one generation to another, but they have to function continuously. So do the centrioles and the achromatic figure which they produce. It is a very great help in tracing two chromosomes when only minor coils are present to have the short one anchored at one end to the achromatic figure and to have the other end labeled with a terminal nucleolus; and to have a long one which possesses these same features plus a lateral nucleolus that clearly differentiates it.

As soon as flagellar band 5 reaches the posterior end of the cell and becomes free of the other bands, it becomes tightly wound again, more turns being put in now than existed before it left the other bands. This is soon followed by constriction of the cytoplasm at the posterior margin of the four flagellar bands that have not moved at all (figs. 37, 79, 80). The constriction continues until the parent cell body is divided transversely, thus producing two cells differing greatly in size: a small posterior daughter with one band (fig. 39) and a large anterior daughter with four bands (fig. 38). It is interesting in this connection to note that in transverse dividing species of *Holomastigotoides* with only two flagellar bands the anterior and posterior daughters are of equal size (fig. 92). Usually the greater the number of flagellar bands, the smaller is the posterior daughter in proportion to the parent cell.

HOLOMASTIGOTOIDES DIVERSA
SPECIES NOVUM

Since several nuclei of this organism are illustrated in this paper and since I am preparing a separate paper dealing largely with the pairing of its chromosomes, a name and brief description should be given it now.

The number of flagellar bands is nearly always eight; occasionally an organism with ten is encountered; whether this represents a variation in number or another species, I am unable to say. Owing to the larger number of flagellar bands, the anterior end is slightly less pointed than that of *H. tusitala,* but the parabasals and axostyles of the two species are identical (fig. 40).

There are forms with 2, 3, and 4 chromosomes (figs. 40–45; 84, 85). All chromosomes have terminal centromeres and, of course, are rod-shaped. In the form with two, one is long and one short; both have terminal nucleoli, and the long one has a lateral nucleolus (fig. 40). This nucleolus, however, is considerably nearer the centromere end of the chromosome than that of the long chromosome in *H. tusitala,* and the chromosomes at metaphase are shorter and broader—because they have fewer and larger major coils—than those of *H. tusitala* (compare figs. 40–45 and 20d–25d). The form with four chromosomes is a diploid, and the forms with three chromosomes may be composed of two short ones and one long one, or two long ones and a short one. In other words, in this species there are haploid and diploid individuals and two types of aneuploid individuals. And if we look at the structure of the chromosomes, each type of individual is divisible into two types, those with single (figs. 40–45) and those with double chromatids (figs. 84, 85). This makes four types of aneuploids, two of haploids, and two of diploids. It is difficult to decide whether these eight types represent species or races. It is certainly more convenient to regard them as races or varieties of a single species. All evidence indicates that each type leads an independent existence. Except for the diploids, which are larger cells, no differences other than chromosomal number and chromosomal structure can be seen.

The centrioles which follow flagellar bands 4 and 5 are shorter than those of *H. tusitala* and do not maintain connection with the nucleus throughout their life cycle. However, they are always present in an elongated condition and produce a new achromatic figure in the late telophase, although, unlike that of *H. tusitala,* it does not function during either the prophase or the resting stage (figs. 40, 44, 45, 84, 85). Connection between achromatic figure and centromeres is not made until metaphase.

Cytoplasmic division is always parallel with the long axis of the body, that is, longitudinal; it is never transverse, as is always true of *H. tusitala.* In fact, division must be longitudinal since both sets of flagellar bands rotate in the unwinding process, just as it must be transverse in *H. tusitala* where only one band rotates. During cytoplasmic division flagellar bands 1, 2, 3, and 4 rotate in one direction and bands 5, 6, 7, and 8 in another direction (fig. 90). This flattens the anterior end of the cell and brings the achromatic figure near the nucleus so that its astral rays may join the centromeres and thus become chromosomal fibres. The nucleus must divide before unwinding of the two sets of bands can progress very far. Since the central spindle connects the two sets of bands (1–4 and 5–8), it sometimes serves as a check, by absorbing some of the tension exerted by the two sets of bands rotating in opposite directions, and prevents the opposite movement of the bands from pulling the nucleus in two prematurely. On the other hand, if the unwinding of the bands is progressing more slowly than usual and thus tends to prevent separation of the two sets of daughter chromosomes, the central spindle has sufficient rigidity to push the bands far enough apart to permit nuclear division to occur at the proper time.

CENTRIOLES AND ACHROMATIC FIGURE

There are so many new, interesting and unusual features in the behavior of the centrioles and the achromatic figure, particularly in transverse dividing species of *Holomastigotoides,* that a brief, separate description of these organelles is essential for a clear understanding of all phases in the life cycle of chromosomes.

In the late telophase of *H. tusitala,* after cytoplasmic division, the centrioles, which are entirely free of centrosomic [3] material, follow flagellar bands 4 and 5 for 1½ turns (text-fig. 2). The two chromosomes, long one (dark) and short one (light), are anchored to the old centriole by means of the remaining portion of the achromatic figure (the chromosomal fibres which extend from centriole to the centromeres). A new flagellar band (no. 5) has grown out. A new centriole has also grown

---

[3] In one species of *Holomastigotoides* in *Psammotermes* the distal ends of the centrioles are surrounded by centrosomes.

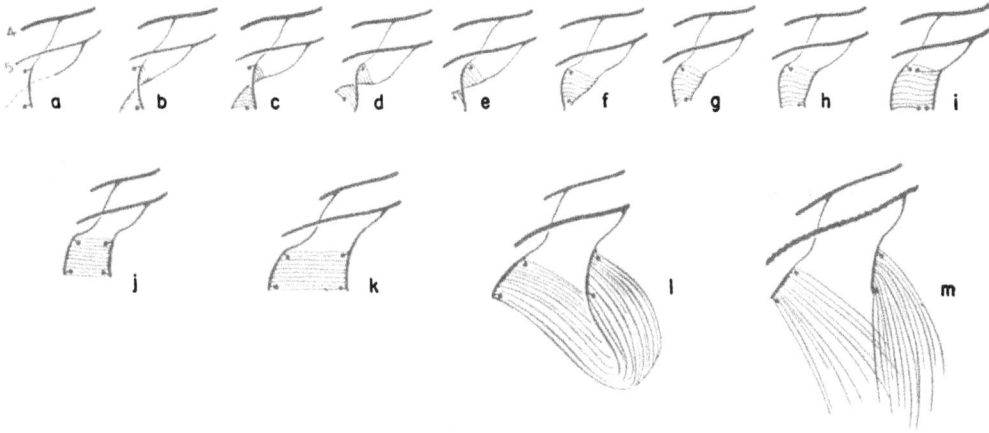

TEXT-FIG. 3. *Holomastigotoides tusitala*. (*a*) Both centromeres attached to centriole that follows old band 4. New centriole has grown out to point where line is broken. (*b*) Astral rays from old and new centrioles about to join to form central spindle of new achromatic figure. (*c*) Central spindle formed. (*d*) Lower right centromere has shifted its connection from old to new centriole. (*d–h*) Centrioles move to parallel positions as achromatic figure develops and central spindle becomes more rigid. (*i–j*) Centromeres duplicated and one from each group moves. (*k–m*) Growth and function of achromatic figure.

out and leaves band 5 after having followed it for 1½ turns. This centriole, like the old one, is free after having followed its band and spiraled the body of the cell for 1½ gyres, but the new centriole is still not quite as long as the old one (broken line, text-fig. 3, *a*, indicates how much it has to increase in length to be as long as the old one, and the course it will take in doing so). When the new centriole has become as long as the old one, or nearly so, both centrioles begin to produce, from their free distal ends, astral rays which are very short and which soon begin to meet, grow along one another and thus form the central spindle portion of the achromatic figure (text-fig. 3, *b, c*). By this time, or thereabouts, an astral ray from the new centriole becomes connected with the centromere of one of the chromosomes and thus becomes a chromosomal fibre. It moves the chromosome to the centriole or pole (text-fig. 3, *d, e;* text-fig. 4, *d, e*). Either the long (dark) or the short (light) chromosome may be moved to the new centriole, but never both; one always stays with the old centriole (text-fig. 4, *d, e*). This, so far as I know, is the first description of the movement of a telophase chromosome, and the first example of the movement of a chromosome (not chromatid) when only two chromosomes are present. It is really an example of the movement, by the achromatic figure, of a single chromosome.

Later, in the very early prophase, as soon as duplication of the chromosomes has occurred and long before the unwinding of the chromatids has made much progress, astral rays connect with the centromere of each chromatid (text-fig. 4, *i*). When enough unwinding (losing of relational coiling) has taken place to permit movement of the ends, two chromatids go to each pole

(centriole). One chromatid is connected to each corner of the achromatic figure which, at this stage in its development, is almost a square (text-fig. 4, *j*). These chromatids for their entire length, except a small portion at their centromere ends, are in the prophase; these small portions are actually in the telophase. This unusual situation has nothing to do with the chromosomes; it merely arises because the achromatic figure is always present and always lies near enough to the centromeres to connect with them when duplication occurs. A close examination of the situation shows that only two of the four centromeres actually move, one of a short (light) chromatid and one of a long (dark) chromatid; the others remain where the chromosomal centromeres were before duplication occurred (text-fig. 4, *h, i, j*).

In the closely related species, *H. diversa*, the two chromosomes lose their connection with the achromatic figure in the telophase shortly after nuclear division and do not regain it until metaphase. An achromatic figure is present, just as in *H. tusitala*, but it does not move the chromosomes because it lies too far away (figs. 40, 44, 45, 84, 85; text-fig. 5). We may conclude from these two types of behavior, which occur again and again in these and related species of *Holomastigotoides*, that chromosomes or chromatids may be moved at any stage in their life cycle when an achromatic figure lies near enough to them for its astral rays to make connection with their centromeres.

Returning to an account of the development of the achromatic figure, we note that as the astral rays, which have joined to form the early central spindle (text-fig. 3, *c, d*), increase in length, the central spindle gradually becomes broader and the distal ends of the centrioles

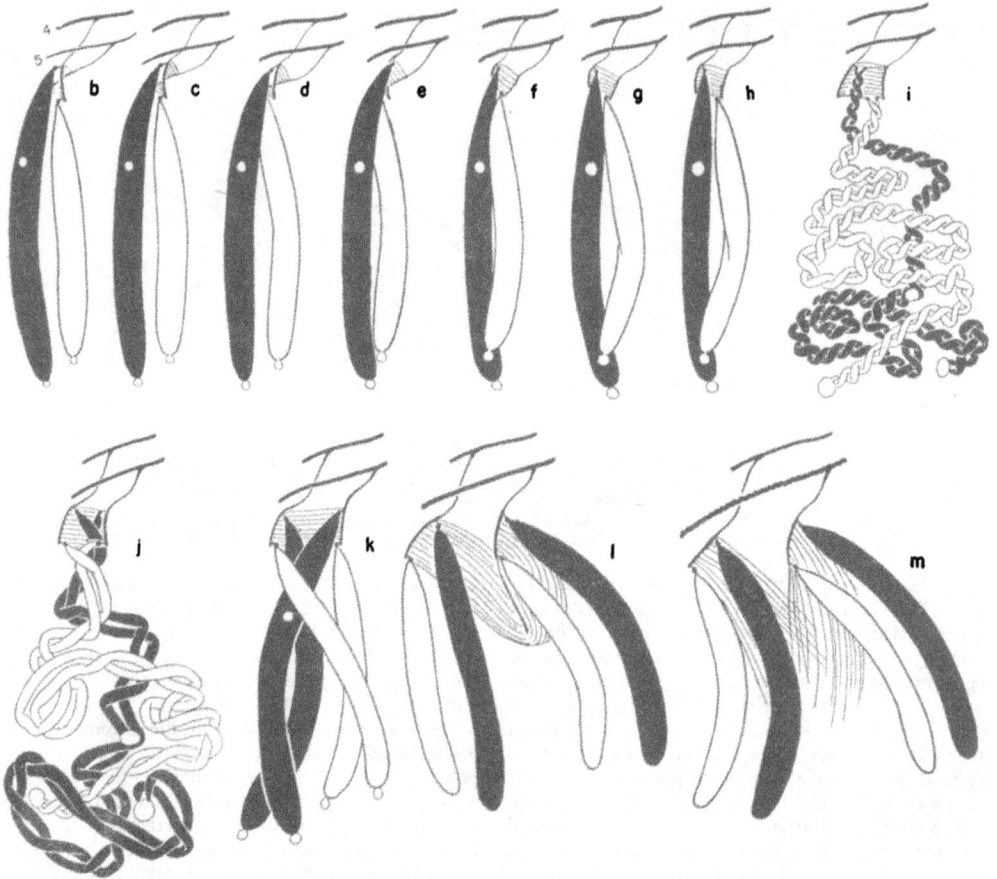

TEXT-FIG. 4. *Holomastigotoides tusitala.* (*b–h*) Telophase; (*i–j*) prophase; (*k*) metaphase; (*l*) anaphase; (*m*) telophase. Brief diagrammatic life cycle of centrioles, centromeres, chromosomes, and achromatic figure. (*b, c*) New and old centrioles forming achromatic figure. (*d*) One chromosome has shifted its connection from old to new centriole. (*e–g*) Flattening out of centrioles and achromatic figure. (*g–h*) Beginning of chromosomal twisting. (*i*) Chromosomes duplicated, producing many gyres of close-together relational coiling of chromatids. Centromeres duplicated; one in each group moves; other remains stationary. (*j*) Later stage in chromatids losing their relational coiling by unwinding. (*k*) Relational coiling practically gone, achromatic figure elongating and separating sister chromatids. (*l*) Central spindle bent, chromatids in two non-sister groups. (*m*) Central spindle pulled apart.

move farther apart until finally they extend posteriorly in the same direction, usually in the same plane (text-fig. 3, *e-j;* text-fig. 4, *e-j*). Shortly beyond this point, development (growth) of the achromatic figure stops and it enters the resting stage which lasts from mid-telophase to very late prophase. When the contraction of the matrix of the chromatids stops in the late prophase, the chromatids enter the resting stage, which may last for days or even weeks (fig. 22*b;* text-fig. 4, slightly earlier than *k*). Most of the organisms in a preparation are in this stage. The resting period, which is longer for the achromatic figure than for the chromatids, ends with the beginning of a second period of growth in the achromatic figure and the formation of individual membranes around each chromatid. The central spindle increases in length and thus moves the chromatids farther apart, the short and the long pair each losing the last vestige of their relational coiling (text-fig. 4, *k, l*). Frequently the central spindle increases in length faster than flagellar band 5 moves away from 4 and thus is forced to bend (text-fig. 4, *l;* figs. 27*d*, 28*d*). Finally it pulls apart (text-fig. 4, *m;* fig. 31*d*), the fifth flagellar band becomes free (fig. 31*d*), the nucleus divides (fig. 32*d*), and the smaller of the daughter nuclei, together with the centriole and the remains of the achromatic figure, go with flagellar band 5 as it unwinds (figs. 33*d*,

35, 76d), becomes free in the posterior end of the cell (figs. 36, 37, 79, 80), which divides into posterior (fig. 39) and anterior (fig. 38) daughters, and as soon as the anterior daughter develops a new flagellar band and centriole the cycle of centrioles and that of achromatic figure are completed.

In the drawings of *H. tusitala* no free astral rays have been drawn, first in order to avoid confusion, and secondly when the cell is stained heavily enough to see them plainly the chromosomes are stained too heavily for their details to be clear. In *H. diversa*, however, the free astral rays are plain when chromosomes are suitably stained for details (figs. 44, 45, 84, 85).

RELATION OF CHROMOSOMES TO NUCLEOLI

In the single chromatid varieties of *Holomastigotoides tusitala* and *H. diversa*, each short chromosome has a terminal nucleolus and each long one has a lateral and a terminal nucleolus. These nucleoli always arise from precisely the same place on the chromosomes and are connected with the chromonema of a chromosome by a fine strand or stem (figs. 2e–4e). They lose connection with the chromatids in the metaphase or early anaphase (figs. 25d, 27d), and usually reappear in the fairly early telophase (fig. 1a). Actually, from the standpoint of the whole life cycle of the chromosome, they are present most of the time (figs. 1d–24d). They are present throughout the stages when the chromonema is losing its major coils (figs. 4d–9d), when only minor coils are present (figs. 10e–15e), and when the chromonema is forming (new) majors (figs. 16d–20d). They are a

great help in distinguishing one chromosome or one chromatid from another when only minor coils are present.

Fusion of these nucleoli is a frequent occurrence. Any one may fuse with another, and sometimes, especially in the telophase when major coils have disappeared or nearly so, all of them fuse to form a single nucleolus (figs. 8e, 10e, 11e). Although they arise from the chromosomes, they are entirely lacking in chromosomal homology as shown by the manner in which they fuse (this subject will be considered in a later paper). They do not form on the terminal end of a chromosome merely because it is terminal. Studies on aberrations, which I shall publish soon, show clearly that free ends do not have the ability to form nucleoli. Sometimes discarded nucleoli remain within the individual chromosomal membrane until new ones are forming and may give the impression that they return to their original positions on the chromosomes or chromatids (fig. 58d).

RELATION OF CHROMOSOMES TO NUCLEAR MEMBRANE

The centromeres are always anchored to the nuclear membrane or a portion of it. In *Holomastigotoides*, *Spirotrichonympha*, and a few other genera of hypermastigotes, the nuclear membrane does not remain intact, as in most protozoa, during nuclear division, nor does it disappear entirely as it is supposed to do in most higher forms. Instead, it does something about halfway between these extremes. In the late prophase of *H. tusitala*, each chromatid is clearly cut off and demarked from the rest of the nucleus and from the other

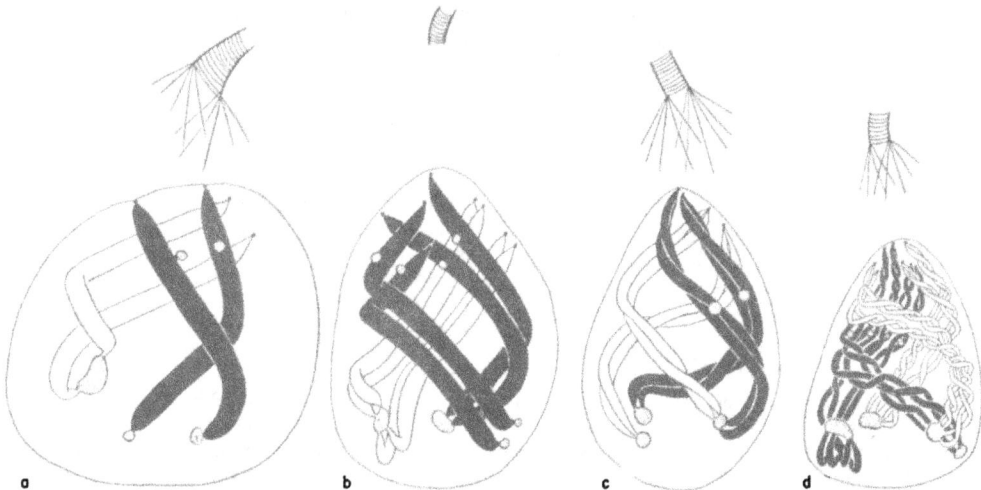

TEXT-FIG. 5. The four types or varieties of *Holomastigotoides diversa*. All are prophases with achromatic figures some distance anterior to the nuclei. The only coiling shown is relational. (a) 2-chromosome form, 4 single chromatids, 4 centromeres. (b) 4-chromosome form, 8 single chromatids, 8 centromeres. (c) 2-chromosome form, 4 double chromatids, 4 centromeres. (d) 4-chromosome form, 8 double chromatids, 8 centromeres.

chromatids by its own membrane (figs. 21e, 69d). Eventually, all the material except that which surrounds each chromatid will be discarded, but in *H. tusitala* the process is very much slower than in *H. diversa* or in *Spirotrichonympha*. The process of discarding a large portion of the old nuclear membrane and its contents is not completed until some time after nuclear division and in the nucleus of what is to be the posterior daughter cell it occurs slightly earlier than in that of the developing anterior daughter (figs. 32e, 33e, 77d). Soon after the old nuclear membrane and its contents have been discarded in the cytoplasm, the two new membranes which surround each chromatid fuse to form a common nuclear membrane (figs. 33e, 79d.) This accounts for the long, narrow nuclear membrane which surrounds the two telophase chromosomes after cell division (fig. 1e). As the major coils relax and thus make the chromonemata appear longer, the nuclear membrane becomes longer (fig. 2e). Then, as the chromonemata bend, the nuclear membrane widens and becomes shorter until the major coils have disappeared (figs. 3e–10e). From this point on, through duplication of the chromosomes, unwinding of chromatids (loss of relational coiling) in prophase, to late prophase the nuclear membrane changes little in shape or size (figs. 11e–20e).

In the double chromatid variety of *H. tusitala*, a new membrane develops around each chromatid rather than each half chromatid (figs. 69d–71d). In *H. diversa* and many speceis of *Spirotrichonympha* where I have studied the formation of a new and the disappearance of an old nuclear membrane in detail, the process begins at early metaphase and is completed by late metaphase or early anaphase. In fact, it is so rapid sometimes that a chromatid becomes free in the cytoplasm before its centromere has made connection with an astral ray. And the amount of nuclear material discarded in the cytoplasm, where it may be observed for several hours, is greater than in *H. tusitala*. Likewise, the new membranes formed around each chromatid are smaller than in *H. tusitala*. This process is directly concerned with pairing of chromatids and of chromosomes. It sometimes plays a decisive role in the type of the pairing that occurs, and this, in turn, with the type of crossing over that results. I plan in a later paper to deal with this situation, together with the possibility that no nuclear membrane may ever disappear completely.

SINGLE CHROMATIDS

The most common type of chromosome in hypermastigote flagellates has a single chromonema in the telophase after cytoplasmic division. There is a second and less common type which, at the same stage in its life cycle, has two relationally coiled chromonemata. To avoid misunderstanding, the two types, for the most part, are treated separately.

In order really to understand chromosomes and their

coiling systems clearly and completely, one must study their entire life cycles.

Previous accounts of mitosis begin with prophase chromosomes which are followed through metaphase, anaphase, and part way into the telophase. Here they stop. This account begins in the telophase near the point where others stop, follows the chromosomes from this point through prophase, metaphase, anaphase, and back to where it began. It shows how the matrix collapses and the chromonemata of the chromosomes lose their major coils (and supers too when present), duplication of chromosomes, origin and loss of relational coiling of chromatids, formation of "new" major and super coils in chromatids, movement of chromatids to the poles, the formation of daughter nuclei, the production of daughter cells, and the existence of the minor coils through all these events.

A chromosome is made up of two very different and equally important constituents, matrix and chromonema. Heavy, intense staining, of any type, usually shows only the sausage-shaped matrix, the old—and very easy—way of seeing and illustrating chromosomes. The chromonema, which is embedded in—not merely surrounded by—the matrix, is not revealed by such technique. Nor is it possible with ordinary microscopic facilities, except in unusually favorable material, to differentiate clearly between matrix and chromonema in the nuclei of living cells. However, with proper fixation and staining technique (see page 2), not only is it possible to see clearly super and major coils of the chromonema, but minors are also quite plain in some types of chromosomes and chromatids, and it may also be seen with equal clarity that each major coil is composed of—has incorporated in it—two or more minors, and that each super has two or more majors incorporated in it. More important still, good phase contrast equipment not only enables one to see all these things in living chromosomes, and thus check on whatever false impressions fixation and staining may have produced, but, in addition, reveals some features, particularly centromeres, minor coils and the demarkation between chromonema and matrix, more plainly than fixed and stained preparations.

The matrix disappears—breaks down—in the telophase and reappears in the very early prophase of each chromosome generation. It may be broken down with weak solutions of ammonia any time and the major and super coils, which were formed by its contraction, freed, just as happens naturally in the late telophase. After duplication of chromosomes, the chromonema of each chromatid becomes embedded in a new matrix. As the matrix contracts, the ever present minor coils are incorporated in a new generation of major coils and at the same time the relational coiling of the chromatids is taken out by the rotation (unwinding) of each chromatid.

It is necessary to distinguish clearly between twisting and coiling. Minor coiling is an expression of the

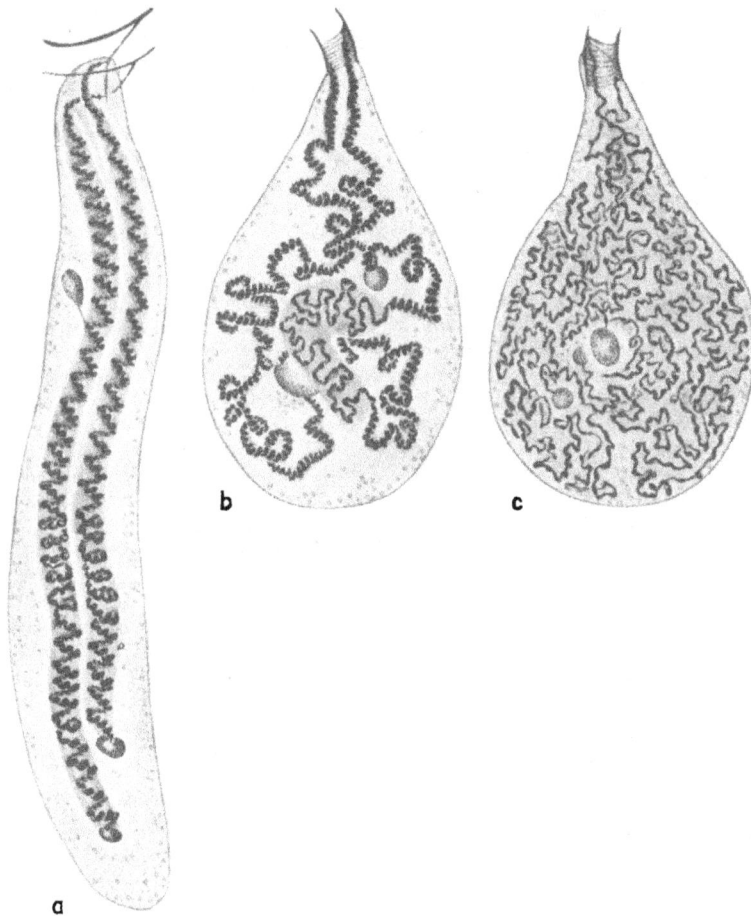

TEXT-FIG. 6. Relation of matrix to chromonema. *Holomastigotoides tusitala*. (*a*) Two single-chromatid, telophase chromosomes after cytoplasmic division. Very early stage in the collapse of the matrix, bringing about a relaxation of the major coils and an apparent lengthening of the chromosomes. (*b*) Late telophase. Varying degrees in collapsing of the matrix. It is completely collapsed in small area in centre and major coils have come out, leaving only minors; in other areas of each chromosome, remnants of major coils remain. (*c*) Very late telophase. Matrix has collapsed completely, its substance is scattered throughout the nucleus where it is disintegrating. All major coils are gone; only minors left.

molecular structure of the chromonema and never disappears; the other types of coils and twisting disappear and reappear during each generation of chromosomes. The next point necessary to an understanding of the coiling systems of chromosomes is a clear conception of the difference between relational coiling and individual coiling. In relational coiling, one chromatid is coiled about another while at the same time each chromatid has its own individual, minor, major or super coils, depending on the type of organism and the stage in the life cycle of the chromatid. Individual coiling may occur either with or without relational coiling, but relational coiling never occurs without individual coiling. In single chromatids, there is no relational coiling from metaphase to very late telophase (figs. 25*d*–39 and 1–10*c*), while in double chromatids there is always one generation (figs. 58*b*–61*b*) of relational coiling, and in part of the life cycle there are two generations (figs. 62*c*–70*d*).

Twisting of chromosomes, which so often has been mistaken for either relational or individual coiling or for both, may, like coiling, be of two types, relational and individual. It is no wonder coiling and twisting have been confused; the only way, in fact, to separate them was to work out the whole life cycle of each. In the telophase as the matrix begins to collapse (text-fig. 6, b) and thus free the gyres of major coils, individual bending and twisting begins in each chromosome (figs. 3a, 3d, 4a, 4d). As more major coils come out, the free ends of the chromonemata sometimes rotate and thus put in relational twisting between the long (dark) and short (light) chromosomes (figs. 5a, 5d; 6a, 6d). Usually not more than one or two gyres of chromosomal twisting are put in, and sometimes only individual twisting occurs (compare figs. 15a and 16a). The relational twisting when present is between non-homologous chromosomes, while relational coiling is always between homologous chromatids or half chromatids. It does not begin to come out until sometime after duplication has occurred and considerable unwinding (loss of relational coiling) in the chromatids has taken place (figs. 19a, 20a). The individual twisting of chromosomes does not all come out until still later (fig. 21a).

In order to follow each type of coiling and twisting through its entire cycle, it has been necessary in H. tusitala to draw most nuclei twice and some of them three, four, and five times. In labeling the drawings, each nucleus, no matter how many times it is drawn, is given the same number, but each drawing of it is given a different letter, and each letter, throughout the plates, refers to the same type of coiling. For example,

a—twisting of chromosome (individual, relational or both)
b—relational coiling of chromatids
c—relational coiling of half chromatids
d—major coils (individual—of chromosomes, chromatids, half chromatids)
e—minor coils (individual—of chromosomes, chromatids, half chromatids).

Super coils when present are labeled. Those numbers without letters refer to drawings of a cell, portions of a cell, and to those nuclei where two or more types of coils are shown in the same nucleus; in some chromosomes or chromatids only minors are drawn; in others majors, supers, and relational coiling are drawn (figs. 46–57, 86).

In following each type of coiling through its life cycle, one soon discovers that the coiling systems are so interrelated that it is impossible to follow or explain one without some discussion of the others at the same time. From late prophase, through metaphase, anaphase, and until what might be appropriately termed mid-telophase, minors are incorporated in fairly tight major coils, from 6 to 8 gyres of them in each gyre of majors (figs. 22e–33e, and 1e–4e). From mid- to late telophase they are in very loose majors (figs. 5e–9e), and, after the majors have disappeared completely, they are free (fig. 10e). After complete disappearance of the matrix, which produced the majors and held them in place until it began to

collapse, the tension on the minors is released, but this does not affect them greatly (text-fig. 6, c; fig. 10e). They are slightly looser—gyres are somewhat farther apart—than after duplication of the chromosome when each chromatid is developing new majors (fig. 16e).

Soon after cytoplasmic division, the major coils become looser and more relaxed (compare figs. 33d and 1d). As relaxation continues, the chromonema becomes apparently longer (fig. 2d) and soon begins to bend (fig. 3d). By this time, it is clearly evident that the matrix is collapsing and that its substance is diffusing through the nucleus (text-fig. 6, b). As collapsing of the matrix continues, the majors become more and more irregular, some of them flatten out, with only a vestige remaining (figs. 4d–6d). This is followed by a complete disappearance of half or more of them, with a few here and there maintaining a semblance of their former appearance (figs. 7d, 8d). Finally, only a remnant of the major coils remains, and this in a portion of a chromosome where the collapsing of the matrix is not complete (fig. 9d). This is followed by the complete disappearance of the major coils— not the slightest remnant or relic of them being left anywhere along either chromosome. Only minor coils (fig. 10e) and chromosomal twisting now remain (fig. 10a). The matrix material in which each chromosome was once embedded has collapsed completely and is now scattered over the entire nucleus (text-fig. 6, c). This material, which presently begins to distintegrate, is probably thrown into the cytoplasm, along with other substances, when the nuclear membrane is discarded and a new one organized (figs. 22e–33e).

Most cells in higher animals and plants and many in protozoa, too, stop at this stage, and there is little activity in the chromonema for varying periods of time, depending on the kind of cell, environment, and other things. This is known as the resting stage or interphase in such cells. But Holomastigotoides does not stop here. Each chromosome soon produces two chromatids, each of which has the same number of minor coils as the parent chromosome which produced it, and one chromatid is tightly coiled (relational coiling) about the other (fig. 11e). Soon after duplication has occurred, each chromatid develops a new matrix in which it is embedded. As this matrix material contracts, the chromatids rotate (unwind) and thus gradually lose their relational coiling (figs. 12e–21e), their apparent length becoming less as contraction progresses. By the time contraction is half completed or a little more and most of the gyres or relational coiling have been lost, the minor coils have become so tight that they are forced to bend in places (fig. 16e); and thus the new generation of major coils first makes its appearance (fig. 16d). So-called "relic coils"— which very definitely do not exist at all—are a result of bridging over the interval between disappearance and reappearance of majors with chromosomal twisting.

As contraction of the matrix progresses and more and more relational coiling is lost, the chromatids appear shorter and their major coils become more regular and hence plainer (figs. 17d–20d). It should be noted that the minors, which were as tight as it is possible for them to be when they began to bend and become incorporated in the majors, do not change in appearance at all as the majors become more regular (figs. 16e–20e). Nothing can be plainer and more certain than the fact that minors do not increase in size to become majors, but instead the chromonema, of which they are a part, merely bends and thus gradually, owing to continued contraction of the matrix, builds up plainer and plainer majors, in which the minors, without any change whatever, are incorporated.

When most of the relational coiling has been lost and the major coils are as tight—close together—as they will ever be, development, so far as the chromosomes of H. tusitala and H. diversa are concerned, stops and the resting stage, which may last for days or even weeks, begins. This is the late prophase and the achromatic figure, which has not increased in size since late telophase, rests as long as the chromosomes do. When the achromatic figure begins to grow again, the two groups of chromatids, with no change in their major coils, are carried to the poles. In fact, there is little, if any, change in these coils during metaphase, anaphase, and early telophase. They do not begin to change until mid-telophase, the point where the discussion of them began.

Not all species of Holomastigotoides behave like H. tusitala and H. diversa; that is, stop and enter the resting stage when major coils become close together. The matrix of many species in the genus continues to contract after the major coils have become as tight as it is possible for them to be, the chromonema is thus forced to bend, and as contraction continues, a new generation of super coils is slowly built up, in which major coils are incorporated in the same manner as minors were incorporated in the majors at an earlier stage (figs. 48, 51). These super coils remain until fairly late telophase (fig. 50), and when the matrix begins to collapse they disappear slightly earlier than the majors. Then, since they come in later and disappear earlier than the majors, they are present for a shorter period in the life cycle of a chromosome or a chromatid than any other type of coil.

I think it should be stated at this point that when double chromatids put in super coils (fig. 86), the two relationally coiled half chromatids, which have individual minors and majors, do not put in individual supers, but instead they put in a common super coiling system. This is interesting because the very same thing occurs in the chromatids of a chromosome at meiosis I, but does not occur in the same chromosome at mitosis. Super coils at meiosis and not at mitosis accounts for the apparent difference in length of meiotic and mitotic chromosomes. Even if meiosis I chromatids had individual centromeres and no relational coiling, they could not go to opposite poles because their common super coiling system would prevent separation until the matrix collapsed, which is in the late telophase, too late for poleward movement.

Relational coiling has been mentioned several times already; now all that remains is a consideration of the manner in which it is produced, when in the life cycle of a chromatid it appears, how long it lasts, and the forces which resolve it.

Nothing can be plainer, if one takes the time to study and consider carefully the coiling systems of chromosomes and chromatids, than the fact that there would be no relational coiling of chromatids if the chromosome which produces the chromatids did not itself have minor coils at the time of duplication (see page 3). The chromosome produces two chromatids, each of which has a minor coiling system exactly like itself. This is the manner in which relational coiling of chromatids is produced and, just as in any two objects coiled about each other, the only way to separate the chromatids is to rotate either one or both of them. If rotation of the chromatids is brought about by movement of their ends without torsion, the unwinding would always have to begin at the ends and proceed along the chromatids. On the other hand, if the contracting matrix produces torsion in the chromonema of each chromatid, then unwinding may begin anywhere, or in several places, along the chromatids. A careful study of chromosomes at the time of duplication and shortly thereafter shows clearly that torsion is responsible for the unwinding of relationally coiled chromatids (see fig. 11e where unwinding at z, a long way from either end, has occurred; figs. 15b–17b where about the same degree of unwinding has occurred all along the chromatids, a situation which would be impossible if unwinding began at one end and progressed toward the other). Proof that the distal ends of the chromatids rotate can be seen when fusion of the terminal nucleoli sometimes retard it. The best examples of this occur in double chromatids, which will be described in the next section (figs. 87b, 88b).

There are only two places in the life cycle of chromosomes and chromatids when rotation occurs. The first is in the late telophase when major coils are coming out due to collapsing of the matrix. The chromosomes rotate and thus produce individual twisting and sometimes a small amount of relational twisting (figs. 3a–21a). But this relational twisting is of chromosomes (not chromatids) and there are never more than two or three gyres of it, while in the early prophase there are over 300 gyres of relational coiling in the long chromosome (11c, 12b). It is thus impossible, after careful consideration, to confuse twisting and relational coiling in Holomastigotoides. The second place where rotation occurs, and this time it is of chromatids, is in the early prophase where, as already explained, it very plainly serves to take out the relational coiling—not to put it in. Then, since neither of these rotations—and there are no more—puts in relational coiling, it must be accounted for on some other basis—and only one is left—duplication of a chromosome when individual minor coils are in.

Details of all the procedures of duplication, of course, are unknown. Perhaps it is sufficient at present to say that a chromosome, like a cell in preparation for division, gradually builds up in duplicate all those substances necessary for continued independent existence and that these substances are slowly segregated into two groups, each of which, physically and chemically, is an exact replica of the chromosome. Duplication, even though we know next to nothing of the processes involved, is at least more accurate from the standpoint of terminology, if for no other reason, than splitting; for, if a chromosome were really to split, as the term implies, two half chromosomes would be produced, not two daughter chromosomes.

DOUBLE CHROMATIDS

The time and manner of disappearance and reappearance of major and super coils is the same irrespective of whether a chromatid is single or double. The main difference in the two types of chromatids is in the time centromeres are duplicated and in the relational coiling; double chromatids always have relational coiling; before one generation disappears (by unwinding), another is produced; so that for part of the life cycle there are two generations. And major coils are smaller in double than in single chromatids (compare figs. 1d and 58d; 44 and 84; 45 and 85; 49 and 86).

I have been unable to think of an appropriate term for each member of a double chromatid and, since the members must be referred to individually many times, a name is necessary. Until such time as someone proposes a desirable name, I am referring to them as half chromatids, although I realize that each half chromatid, except for lack of its own centromere, is identical with a chromatid. Actually, these so-called half chromatids are structurally chromosomes, and we should develop a terminology that would permit us to call them chromosomes. After cell division they will be called chromatids, and in the next generation, after cell division, they will be chromosomes.

Each half chromatid has a terminal nucleolus, and the long ones in addition have lateral nucleoli. Frequently the terminal nucleoli of chromatids are fused (figs. 58b, 59b, 65b) and those of the half chromatids are nearly always fused. The period when this nucleolus cannot be seen is very brief. It can be seen in some cells in all stages, while in others it cannot be seen from late prophase to mid-telophase. The fact that, even when these nucleoli cannot be seen, the distal ends of the half chromatids are nearly always joined suggests their presence (figs. 70d–80d). New ones are probably formed immediately after the old ones lose connection with the half chromatids and fusion follows so quickly that it is rare to see free ends of half chromatids either with or without

nucleoli (figs. 72d, 73d). Either the time when old nucleoli disappear and new ones are formed varies considerably or else the old ones sometimes persist for a fairly long time after losing connection with the half chromatids (fig. 80d).

Each of the two telophase chromosomes (fig. 58) of H. tusitala, after cell division, is composed of two relationally coiled chromatids (fig. 58b), each of which has individual major (fig. 58d) and minor (58e) coils. These chromatids, if not relationally coiled, might be mistaken for individual arms of two-armed chromosomes. Some individuals to whom I have shown them have suggested this possibility, even in spite of the relational coiling. But, if one follows their subsequent development, it is perfectly clear that they are chromatids and not chromosomal arms.

The individual membranes around each telophase chromosome fuse completely; the chromosomes become apparently longer and begin to twist (fig. 59a). At the same time the relational chromatid coiling (fig. 59b) and the major coils (59d) become looser. As collapsing of the matrix continues, the chromosomes twist more (fig. 60c) and the majors become much looser (fig. 60d). By this time the relational coiling of the chromatids has become quite loose, but none of it has come out. This occurs only when the matrix contracts.

Meanwhile the new flagellar band and centriole grow out and the centrioles form a new achromatic figure. Before the achromatic figure was formed, the two double-chromatid chromosomes were anchored to the old centriole by the chromosomal fibres of the old achromatic figure; one chromosome being anchored to the upper and one to the lower portion (figs. 58b, 59b). Now, when the new achromatic figure is formed, instead of one chromosome remaining anchored to the old centriole and one moving over and becoming anchored to the new one, as occurs in the same stage in the life cycle of single-chromatid chromosomes (text-fig. 4, b–f, fig. 1a), one chromatid moves to the upper and one to the lower portion of the new centriole, so that there are four centromeres anchored to the achromatic figure, one being anchored to each of the four corners (fig. 60b). At this same stage in the life cycle of single-chromatid chromosomes, only two centromeres are anchored to the achromatic figure (figs. 4d–6d). They are not anchored to opposite sides; that of the long (dark) chromosome is anchored to the lower right (figs. 4d, 6d). They may be anchored in the opposite manner, namely that of the long one upper right corner and that of the short one lower left (fig. 5d). As already explained, the manner in which they are anchored to the new achromatic figure depends on how they were anchored to the old achromatic figure and which chromosome moved when the new figure was formed.

This very striking difference in the behavior of the centromeres of single- and double-chromatid chromosomes deserves further consideration. In the single chromatid variety, the centromeres separate immediately after duplication occurs, even though over 300 gyres of very tight relational coiling are present (fig. 11e). One centromere of each chromatid remains anchored to the same centriole that the chromosomal centromere was anchored to prior to duplication; the other moves (compare figs. 10e and 11e).

Now, when the chromatids of double-chromatid chromosomes are duplicated, the half chromatids do not get individual centromeres (fig. 62c). At least if they do, the two centromeres do not separate, for sister half chromatids certainly appear to have a common centromere. Failure of these centromeres to separate cannot be due to the presence of relational coiling, since they do not separate in later prophase when only a few gyres are left (figs. 67c, 68c). In fact, so far as one can see, the sister half chromatids have a common centromere through metaphase, anaphase, and until very late telophase (fig. 60b). However, if duplication of the centromeres occurs after the two poles of the achromatic figure have moved a considerable distance apart (figs. 73c and later), or after they are disconnected, failure of the centromeres to go to the poles cannot be used as evidence that duplication has not occurred, because one pole, or two disconnected poles, will not move centromeres and their chromatids. This statement is based on observations of a large number of really true unipolar achromatic figures, in which no poleward movement whatever of centromeres or chromatids occurs. Astral rays from a single pole may touch centromeres, but contact is never established.

However, it is evident from observations on the single chromatid variety of *H. diversa* where the astral rays of the achromatic figure are not near enough to the nucleus to make contact with the centromeres until metaphase, that centromeres can and do move apart as soon as duplication of the chromosomes occurs, long before the achromatic figure is near the nucleus (fig. 45). This is true in both haploids and diploids (figs. 44, 45). But in the very same stage in the life cycle of double-chromatid chromosomes, the sister half chromatids have common centromeres. This is true both in haploids (fig. 84) and in diploids (fig. 85). And in later stages when the achromatic figure has made contact with the nucleus and the nuclear membrane is disappearing, these sister half chromatids still have common centromeres. In fact, they keep them through anaphase and until late telophase, just as in the double chromatid variety of *H. tusitala*

Now, since it is perfectly clear that the centromeres of chromatids can, and in fact do, separate whether an achromatic figure is present or not, we must conclude that failure of the centromeres of half chromatids to separate means they have not been duplicated. Chromatid duplication produces half chromatids in the very early prophase, but centromere duplication in half chromatids does not occur until late telophase.

The behavior of the centromeres of single-chromatid chromosomes, as described here, is the same as in mitosis —haploid or diploid, protozoon, higher animals or plants —and the behavior of the centromeres of double-chromatid chromosomes is the same as in meiosis. Except in a few primitive types of meiosis, which I am describing in later papers, each metaphase I chromosome is composed of two relationally coiled sister chromatids with a common centromere (in some species, each chromatid is composed of two relationally coiled half chromatids, but this does not alter the comparison being made). Now, in the usual types of mitosis, diploid or haploid, the metaphase sister chromatids are not relationally coiled and they do not have a common centromere. Each chromatid has its own centromere. Is it possible in meiosis I that the failure of the centromere to duplicate itself when the chromosome is duplicated is responsible for the retention of the relational coiling? If so, the centromere plays a very important role in mitosis and meiosis, especially in changing from one to the other. In fact, a change in its behavior may be necessary before a cell, by gametic meiosis, can produce gametes. And in zygotic meiosis, a change in its behavior may be necessary before a cell, following fertilization, can return to the haploid condition.

There are two cardinal differences between mitosis and meiosis.[4] They concern the duplication of chromosomes and centromeres. In mitosis, both chromosomes and centromeres are duplicated at the same time. In the prophase of meiosis I, the chromosomes are duplicated; the centromeres are not duplicated. They are duplicated either in late telophase I or early prophase II. Then, meiosis I, so far as the number of centromeres is concerned, is a haploid—there is one centromere for each chromosome, not one for each chromatid, as in mitosis. In prophase II, the chromosomes are not duplicated; all that happens between meiosis I and II is for the chromatids of I to lose their relational coiling. This, then, brings about a reduction in the number of chromosomes. In meiosis II, the chromosomes and centromeres are both haploid, and this division is like haploid mitosis. Then, in meiosis I a generation of centromeres is lost and in meiosis II a generation of chromosomes is lost. Both of these are regained when fertilization occurs.

Now, any environmental change which will suppress first a generation of centromeres and then a generation of chromosomes, can produce meiosis, provided, of course, the centrioles and other organelles concerned in cell division continue to function normally. However, I should point out that such an environment may not in itself always produce gametogenesis. Other factors

[4] I am thinking here of the usual type of meiosis which requires two divisions for completion. In later papers I will describe other types of meioses: one in which the duplication of both centromeres and chromosomes is inhibited during the first nuclear division; another in which the duplication of neither centromeres nor chromosomes is inhibited in the first nuclear division, but both are inhibited in the second.

may be concerned in this process. More experiments are necessary before this question can be answered. For example, in *Barbulanympha* meiosis occurs even when no gametes are produced (Cleveland, 1947b). In *Trichonympha* of the roach *Cryptocercus*, both gametogenesis and meiosis occur, but not at the same time. Fertilization is between gametogenesis and meiosis. The first division following fertilization is meiosis I. In *Trichonympha*, experiments (to be reported in other papers) show that the molting hormone produces gametogenesis, but it may not be responsible for meiosis, since both fertilization and zygotic meiosis occur when the cysts containing the gametes are transferred to non-molting roach hosts. However, the hormone which converted asexual cells into gametes may persist in the gametes and in the zygote. If so, then, it is also responsible for meiosis. On the other hand, it may be that in *Trichonympha* once gametes have been produced, they must fuse or die, and that after fertilization has been accomplished meiosis must follow—with no special environment or substance being necessary to make the centromeres and chromosomes behave as they do—but I doubt it very much since the varied assortment of genera in *Cryptocercus* all undergo meiosis at the same time, the behavior in those genera where no gametes are produced coinciding precisely with that of those genera which produce gametes.

This diversion, I hope, has prepared the reader for an explanation I wish to suggest regarding the possible origin of the similarity in behavior of the centromeres of double-chromatid chromosomes in each and every cell division with those of most organisms at meiosis I.

All the protozoa (12 genera, 7 families) of *Cryptocercus* have sex of one type or another. *Holomastigotoides* and several other genera of the Spirotrichonymphidae of termites where double-chromatid chromosomes have been studied are closely related to the protozoa now living in *Cryptocercus*. No sex exists in them now, but it very probably existed earlier and has been lost (see p. 5). If so, perhaps we are observing a remnant of it. The likelihood of such a possibility does not seem too remote if one considers the fact that very slight changes are necessary to convert meiotic processes into the same type of behavior as occurs in these double-chromatid chromosomes and their centromeres. All that is needed to change to this type of behavior is for the chromosomes to duplicate themselves between meiosis I and II, something they do not do in meiosis, and for the centromeres to continue to behave as in meiosis. These forms have centromeres which, from the standpoint of time of duplication, behave as in meiosis; and they have chromatids which, from the standpoint of time of duplication, behave as in mitosis. In other words, this seemingly unusual situation may result from the retention of one cardinal meiotic form of behavior and the loss of another.

In *H. diversa*, for example, the two double-chromatid varieties (text-fig. 5, *c, d*) which are labeled haploid and diploid (figs. 84, 85), are haploid and diploid only from the standpoint of their centromeres; from the standpoint of their prophase chromatids, they are diploid and tetraploid. However, the two single-chromatid varieties of *H. diversa*, are true haploids and diploids, both from the standpoint of the centromeres and the prophase chromatids (figs. 44, 45; text-fig. 5).

Returning to the life cycle of the double-chromatid chromosomes, we note that the collapse of the matrix soon progresses so far that only the barest remains of the major coils are left (fig. 61d). This is followed by complete disappearance of the major coils, much chromosomal twisting (fig. 62a), and duplication of the chromatids, each producing two relationally coiled half chromatids (fig. 62c) each of which has its own individual minor coils (fig. 62e). The old generation of relational coiling, which began in the previous prophase, persisted through metaphase, anaphase and telophase, is still present (fig. 62b), so that from this stage (very early prophase) in the life cycle on, there are two generations of relational coiling, that of chromatids and of half chromatids.

Now each half chromatid develops a new matrix in which it is embedded. When these new matrices begin to contract, the minor coils of the half chromatids become tighter, the half chromatids rotate and thus begin to lose their relational coiling (fig. 63c). This is the new generation of relational coiling, and as rotation of the half chromatids removes it, rotation of the chromatids continues to remove the old or previous generation of relational coiling of chromatids (fig. 63b). At the same time chromosomal twisting, which may be only individual (fig. 63a) or both relational and individual (fig. 64a), begins to unwind. From this point on the old and new generations of relational coiling and chromosomal twisting all unwind at the same time, but only the chromosomal twisting and the old generation of relational coiling come out completely in this generation.

Unwinding of the relationally coiled half chromatids does not progress very far before the minor coils become so tight that the chromonema begins to bend; as the bending progresses, the minors are incorporated in a new generation of small, new, loose majors (fig. 64d). By this time, considerable progress has been made in the unwinding of the old generation of relational coiling (fig. 64b), only a few gyres being left, while the new generation (relational coiling of half chromatids) has many gyres left (fig. 64c).

Presently the gyres of major coils in the new generation become close enough together that they may be recognized easily (fig. 65d), while slightly over one gyre of relational chromosomal twisting (fig. 65a), approximately two gyres of relational coiling of chromatids (fig. 65b), and many gyres of relational coiling of half chromatids remain (fig. 65c).

In slightly later stages, where more contraction of the matrix has occurred, one can now see that the chromosomal twisting (fig. 66a) and the old generation of relational coiling (of chromatids) are really coming out (fig. 66b). Approximately a gyre of each remains. The major coils are considerably tighter (fig. 66d) and many gyres of relational coiling of half chromatids remain (fig. 66c).

But neither the chromosomal twisting nor any of the coiling systems come out completely before all unwinding comes to a complete standstill and the resting stage is entered (fig. 67a-d). In this stage, which all observations indicate may last from several days to weeks, there is usually less than one gyre of individual chromosomal twisting, and if relational twisting of chromosomes exists at all there is less than a gyre of it (fig. 67a). As a rule there is also approximately one gyre of relational coiling of chromatids at this time (fig. 67b), while there may be as few as two or as many as 12 gyres of relational coiling of half chromatids, the usual number being 5 or 6 (fig. 67c). The major coils are close together but not as close as in later prophase (figs. 68d, 69d, 70d), metaphase (fig. 71d) and for a long time in the telophase (figs. 72d-79d).

In that part of the prophase development following the resting stage, relational twisting of chromosomes (fig. 68a) and relational coiling of chromatids (fig. 68b) begin to disappear and soon both are lost completely. From this point on until late telophase, when rotation of chromosomes results from the collapsing of the matrix, there is no twisting of chromosomes. This, then, takes the twisting to the point where the account of it began. And from prophase development following the resting stage, there is only one generation of relational coiling (that of the half chromatids) until after chromosomal twisting has returned: in fact, the next or new generation of relational coiling of half chromatids does not appear until duplication of the chromatids (fig. 62c). Of course, after cell division, what were previously half chromatids become chromatids.

From very late prophase, metaphase, anaphase, and to fairly late telophase the relational coiling of half chromatids does not come out at all, and for a very good reason; there is no contraction of the matrix during this period so that no rotation occurs (figs. 68c-78c). We may conclude, then, that relational coiling does not come out except when the matrix is contracting; and that relational twisting and individual twisting do not come in except when the matrix is collapsing.

DISCUSSION

So many papers have been written dealing with chromosomes and their coils and so much difference of opinion exists concerning them that a detailed review of the subject would require over a hundred pages and hence would be out of place here. It is a job for a text-book.

However, it does seem desirable to discusss briefly a few points, particularly centromere, relic coiling, and relational coiling. Aside from this, I am content to present a straightforward and what I believe to be an accurate account of the series of events in the life cycle of the chromosomes of the hypermastigote flagellates I have studied, with no effort whatever on my part to make the observations of other workers on other types of material fall into line with mine, although I myself am firmly convinced, from my own observations on other chromosomes, that they will eventually do so. More studies and less discussion and argument are needed.

During the course of this investigation several animal and plant cytologists have viewed my preparations of these chromosomes of flagellates and without exception they have remarked how closely these chromosomes resemble the known stages in the life cycle of the chromosomes of higher plants and animals. The usual remark, after studying them and comparing them with chromosomes of higher forms, has been "But they are much plainer. This, plus the fact that there are only two and that they are rod-shaped, should enable you to learn many things about them." Not once has anyone who has seen them even suggested the possibility that the fact that they are chromosomes of unicellular organisms might in any way disqualify them as material for generalizations regarding chromosomes.

The basic pattern for chromosome morphology was laid down long ago in unicellular forms and there has not been much evolution in it as higher forms have developed. Rather, the evolution has been in the content —the chemistry—of the chromosomes. For example, there has been no evolution, even in the most minor details, of chromosome morphology in flagellates of the genus *Holomastigotoides* that live in termites of the genus *Prorhinotermes*. These termites occur in such widely separated points as Florida, Madagascar, and Indonesia. Some of them have been isolated since the Tertiary, if not longer, and yet the chromosomes of their flagellates of the genus *Holomastigotoides* are precisely alike. There has probably been no evolution in the chemistry of these chromosomes during this period since all species in all of these termites are identical regardless of where they are found. Other examples similar to this one have been observed.

Several cytologists who have studied my preparations of the centromeres and their relation to the chromonemata have remarked that "Nothing could be plainer than the fact that the centromeres in some genera are all terminal." They have said, "When you write up these studies be sure and stress this fact." But how can I do more than state and illustrate the fact. If, after I have done this, there are still doubting Thomases, let them prepare and study the chromosomes of these organisms themselves.

Those genera with two chromosomes comprise four categories, with several genera in each: in the first, both centromeres are terminal; in the second, one is terminal and one median; in the third, one is subterminal and one median; in the fourth, both are median. In diploid forms of these same organisms, the situation of course is the same except that there are twice as many chromosomes. In *Trichonympha* from *Cryptocercus* the haploid number of chromosomes is 24 and they all have terminal centromeres. In *Trichonympha* from termites all species studied have terminal centromeres. In *Barbulanympha, Pseudotrichonympha, Rhynchonympha, Urinympha, Oxymonas, Pyrsonympha, Saccinobaculus,* and several other genera, all of which have a fairly large number of chromosomes, some of the centromeres are subterminal and some are median. In some of these genera, and in others not mentioned, there is a great difference in the size of the chromosomes. In all, I think one is amply justified in concluding that the chromosomes of at least some unicellular forms present as great a diversity of size and of position of the centromeres as is found in any of the higher forms of life.

The actual relation of centromere to chromosome is very plain when studied in living material with dark phase contrast equipment. The chromonema is fairly dark, almost as dark as when stained with hematoxylin, and the centromere is brown. Under these conditons, the contrast between the two organelles enables one to see both of them clearly, even when the centromere lies adjacent to and sometimes almost seems to be embedded in the chromonema, as in certain plant chromosomes. This brings up a point which I think few chromosome cytologists fully appreciate, namely that the centromeres of the chromosomes of some flagellates lie a considerable distance from the chromonema. In some genera they may even lie 5–10 microns from the chromonema with which they are connected by a strand which, when viewed in living material, is seen to be very elastic. Pressure on the cover glass will increase the distance between centromere and chromonema two to three times without breaking the strand. Eventually, however, the strand will break after being stretched many times, and it is thus possible to separate centromere and chromonema, leaving both intact. Permanent preparations of this type may be obtained by using drastic smearing methods, including sufficient pressure on cells to disrupt their nuclei.

The clear-cut difference in living material between centromere and chromonema shows that the centromere, even though in mitosis of single-chromatid chromosomes it is duplicated at the same time as the chromonema, is more than merely a place on the chromonema where a chromosomal fibre attaches itself. It indicates that the two organelles are probably of very different composition. The same is also indicated by the fact that in meiosis I chromonemata are duplicated and centromeres are not duplicated, while in meiosis II the situation is the other way around. This behavioral difference shows that the centromere is a very important organelle, that

it not only plays an important role in the poleward movement of chromosomes, but is also a key mechanism in chromosomal reduction.

Schrader (1936, 1939), Schaede (1936) and others have shown very clearly that centromere and chromonema do not always stain alike. Their observations together with mine on living material furnish sufficient proof of the individuality of the centromere. In certain species of Heteroptera and Homoptera Schrader (1935), Hughes-Schrader and Ris (1941) have reported diffuse or polycentric centromeres, but Geitler (1938) may be correct in not wishing to conclude, until more work is done, that the centromeres of one or more orders of insects are greatly different from those in other animals and plants.

But whatever else the centromeres may do, they certainly do not form any part of the achromatic figure in any of the organisms I have studied, and I doubt very much that they do in any cells. I cannot agree with White (1945: 18) when he says "At mitosis the centromeres of all of the chromosomes appear to co-operate in producing the spindle, and may be regarded as organizers of the gelation process which converts the fluid nuclear sap and surrounding cytoplasm into the rigid body to which the chromosomes are attached." I really cannot understand why White made this statement, even though he qualified it, for later in the same paragraph he says, "Fragments of chromosomes which have lost their centromeres usually do not become attached to the spindle, but float about freely in the cytoplasm." And later he says that under special circumstances spindles can be formed without chromosomes being present. He certainly has not considered the large number of flagellates where very large achromatic figures are always formed outside the nuclear membrane which remains intact throughout cell division. It is in such organisms that one can see more clearly than anywhere else the role of the centromeres in mitosis and meiosis. The centromere serves as a point in the nuclear membrane where chromosome and chromosomal fibre make contact.

White, however, is not alone in thinking centromeres may sometimes produce the spindle. Darlington (1937: 544) says,

In the cleavage divisions of the egg, spindles are developed by the centrosomes and by the centromeres, independently and simultaneously. These spindles combine and their combination produces metaphase. At meiosis on the other hand the centrosomes play no part at all in the formation of the spindle. It arises entirely within the nucleus, presumably as a centromere spindle. But before the spindle develops the nucleus contracts to a small proportion of its previous size. Now, so long as a nucleo-cytoplasmic surface persists, the difference of structure and behaviour on the two sides of it shows it to be acting as a type of semipermeable membrane. The sudden contraction of the nucleus, therefore, probably means a loss of water and an increase in the contraction of whatever large molecules are present in the nuclear sap from which the spindle develops.

Speculations will never yield definite information regarding the origin of the achromatic figure, nor will they tell us how it functions.

Perhaps I should mention here, since they are pertinent, observations which I plan to describe in detail later. The molting hormone of the roach *Cryptocercus* makes the chromosomes of the flagellate *Trichonympha*, which lives in it, more susceptible to oxygen poisoning than the other constituents of the cell. By treating the roach with 80 per cent oxygen shortly before it molts, it is possible, depending on the length of the treatment, (1) to partially destroy the chromosomes of the gametocyte of *Trichonympha*, (2) to destroy some of them, (3) to destroy all of them leaving the nuclear membrane intact, (4) to destroy all of them together with the nuclear membrane. Even when the last vestige of chromosomes and nuclear membrane are destroyed, the centrioles, which are large enough to be seen easily, are not affected. They produce a perfectly normal achromatic figure, as well as flagella and parabasals, and the cell divides, producing two anucleate gametes.

Further, when the flagellate *Barbulanympha*, which also lives in the roach *Cryptocercus,* is subjected to very low concentrations of oxygen shortly after the roach molts, its centrioles, which are from 28 to 35 microns in length, are destroyed in some cells while in others they are prevented from functioning. As a result of this treatment, no achromatic figure is formed, the chromosomes are duplicated, but nuclear division does not occur. It seems to me that these clear-cut experiments, even though so far they have been carried out on only two genera of flagellates, yield more dependable evidence regarding the function of centromere and centriole than a thousand observations on material where cytologists differ considerably regarding what they see and greatly regarding their interpretations of it.

The failure of the centriole to function in the production of an achromatic figure, due to some change in the environment of a cell, is the probable explanation of endomitosis or endopolyploidy, which has been found recently in so many types of animal and plant cells. The stimulus which inhibits the function of the centriole does not stop the duplication of chromosomes. Usually in *Barbulanympha* three to four days before its host *Cryptocercus* molts, the centrioles are not merely prevented from functioning, but one generation is actually lost. They disintegrate, but before doing so they produce two new ones which replace them. But, in the meantime, the chromosomes divide so that the haploid cell now becomes a diploid. This is the simplest type of endopolyploidy, for, by the time the next duplication of the chromosomes occurs, the new centrioles are ready to function and nuclear division occurs. This behavior is clearly due to the action of the molting hormone on the centrioles. It is highly probable, in my opinion, that the endopolyploidy so common in insect cells of many types, but best known in those of the salivary glands of the Diptera, where multiple chromosomes are produced, is caused by the hormone which is responsible for the successive series of larval stages through which the insect passes and that the profound changes of metamorphosis, resulting in the destruction and replacement of much tissue, are the insect's method of disposing of tissue which, following each progressive larval stage, attains a higher degree of polyploidy and hence eventually becomes useless.

I have used minor, major, super, and relational coiling in the same sense that Darlington and most chromosome cytologists use them, but I have not used the term "internal" coiling. Instead, I have used "individual" coiling because this seems to me a better term. All coiling is of two types: individual and relational. One can speak of individual minor, major, and super coils and of relational coiling. The individual coiling may be of chromosomes, chromatids, and half chromatids, depending on the stage in the life cycle and whether chromatids are of the double or single type. And the relational coiling may be either of chromatids or half chromatids, depending on the type of chromatid, but not of chromosomes. This is twisting, not coiling. I make a clear distinction between the two. Relational coiling results from the duplication of a chromosome or a chromatid when minor coils are present; relational twisting results when two chromosomes which have been separated become twisted about each other.

A chromosome may also have individual twisting as well as coiling, and this has no relationship to individual minor, major, and super coils. If I understand Darlington correctly, he has called this individual twisting "relic coiling" or a "relic spiral." He (1937: 33) says, "At telophase a spiral becomes indistinguishable, corresponding to the supposed structure at metaphase. This is the relic spiral." Then on page 34 he says, "The coiling seen in a prophase is a relic of that at the preceding division, not the forerunner of that at the following division, as has been generally imagined." And on page 490 he says,

The reason why the relic spiral cannot be used in forming the new internal spiral (= individual coiling according to my terminology) is, on the other hand fairly clear: it has already expanded to an amplitude far greater than an internal spiral ever assumes, and an internal spiral always develops by increasing its amplitude, not diminishing it.

All of which is perfectly true, and I think Darlington will agree with me that twisting should be substituted for his terms relic coil and relic spiral if I can convince him that major coils do not merely become "indistinguishable" in the telophase, but that they disappear, leaving behind no remnant or relic whatever. The term twisting is certainly less likely to confuse the student and since, as Darlington says, relic coiling is unrelated to individual (= internal according to Darlington's terminology) or relational coiling, it should not be referred to as a coil, for it is not a relic of any type of coiling—minor, major, super, or relational. Rather, it is a bending and twisting of the chromonema which occurs in the fairly late telophase as the chromosomal matrix begins to collapse and thus gradually frees the major coils which eventually disappear entirely (figs. 6d–10a). I wish to emphasize the fact that nothing can be plainer in the material I have studied than the complete disappearance of all major coils in the late telophase of each nuclear cycle. In the mid-prophase or thereabouts, the place where Darlington first sees his relic coils after losing them (as major coils) in the previous telophase, the individual chromosomal twisting (fig. 12a) does resemble disorganized major coils more than it does in the interphase or apparently longest stage of the chromonema (fig. 10a). This is due to the fact that the apparent shortening of the chromonema preparatory to the development of the new generation of major coils, although hitherto not recognized as such, has already progressed far enough by this time to have tightened in places the looser bends and twists that were present in earlier stages. In fact, if the mid-prophase were the only place in these chromosomes of flagellates where I could see these bends and twists of the chromonema, I very probably would consider them relics or remnants of the telophase major coils, but it is perfectly clear, in chromosomes which may be followed through their entire life cycle, that major coils leave behind no relic.

As development of the new major coils progresses, the twisting of the chromonema comes out, all of it finally coming out by the time the new generation of majors is fairly well organized and, just as Darlington says, it does not enter directly into the formation of the new majors. Rather, the chromonema loses its prophase kinks, bends, and twists—Darlington's relic coils—as it becomes recoiled into a new generation of majors.

But the minor coils which are present at this time—in fact are always present—are not taken out as the new majors come in. Instead they are incorporated in the majors where they lose none of their identity. Likewise, later, in those chromosomes where super coils occur, the majors are incorporated in supers and without losing their identity.

Sparrow (1942) has given the most complete and most accurate account of prophase major coils. He has shown that as the gyres become larger in diameter they become fewer in number. And he has also shown that relational coiling is resolved by unwinding. But he has not shown, nor has anyone else, how relational coiling is produced, how major coils disappear and reappear, and the relation of the ever present minors to the majors. He has shown majors at metaphase, but he has not shown the minors incorporated within them at this or any stage.

I believe that in the course of time the eye experienced in detecting minor coils will be able to find them incorporated in all majors, in mitosis as well as meiosis; that major coils occur in mitosis as well as meiosis, and that the term "standard coil" of mitosis should be dropped; that the apparent lengthening of chromatids at anaphase and telophase results from a gradual loss of the regularity and compactness (close together) of the major coils. This process begins earlier in some types of material than in others, and it even progresses faster sometimes in one chromatid or chromosome than in another in the same nucleus.

If majors disappear completely—and I know they do in the material I have studied—the question of reversal of major coils has no significance, because the new generation of majors would be free to coil in either direction. Further, since it is perfectly plain that the development of new majors may begin in two or more places in a chromatid or half chromatid before they begin all along it, we thus see clearly how differences in direction may occur in the same arm.

The hypothesis of Darlington (1937) that relational coiling is produced by torsion soon breaks down when one applies it to actual conditions observed in chromatids. Instead of coming in as the matrix contracts and the chromatids become apparently shorter (major coils become larger and closer together), relational coiling comes out. In the very earliest prophase, there are many more gyres of relational coiling than in mid-prophase, and there are more at mid-prophase than later. As Sparrow (1942: 264) has pointed out, when first observed, there are so many gyres of relational coiling that they could not possibly have been produced by one chromatid being twisted about another due to torsion. Also, the relational coiling at a fairly early stage is far too regular to have been produced by one chromatid twisting about another. In fact, to produce this regularity both chromatids have to twist, and the twist in each has not only to be in the same direction but at the same rate. Clearly an impossibility.

On the other hand, if one forms no hypothesis at all but merely accepts as fact what he can see and can verify over and over again by direct observation on stained and living material, the manner in which relational coiling is produced is simple and no obstacles are encountered in relating this information to the loss of relational coiling and to the formation and disappearance of major coils. In the very latest telophase of single chromatid chromosomes (fig. 10e), one can see only minor coils; in the very early prophase (fig. 11e), which, in the material illustrated, follows the telophase with no intervening phase or stage, one can see regions along each of the two chromosomes where only a minor coil is present (x), regions where the gyres of relational coiling are very small and close together (y), and regions where some unwinding of the relational coiling has occurred (z). In the first region, one sees minor coils of a chromosome; in the second and third, one sees minor coils of chromatids. At duplication, each chromosome passes on to each of its daughter chromosomes or chromatids an exact replica of its own minor coils. This is brought about by a duplication of the chromosome in the same direction as that of the gyres of its minor coils. Verification of this, in addition to what one can see, may be had by comparing the number of gyres of minor coils in a chromosome before duplication with the number in each chromatid after duplication. One can also see a relationship between minor and relational coiling by comparing the number of minor gyres in a chromosome before duplication with the number of gyres of relational coiling after duplication.

I have seen many times the two arms of a single chromosome, or two portions of the same arm, twisted about one another as a result of the rotation of the chromonema during the disappearance of major coils, but this twisting, which never comprises more than a few gyres, should not, in any way, be confused with relational coiling.

Also, when two chromosomes—it does not matter whether they are the single or double chromatid type—pair in the prophase, they sometimes twist about one another, especially if pairing begins when the majors are still fairly loose or, in those instances where supers are present, before supers are formed. This twisting, which never comprises more than a few gyres and is due to a difference in rate of development of the major (or super) coils in the two pairing chromosomes, should not be confused with relational coiling. It also should not be confused with the relational twisting of non-homologous chromosomes that sometimes occurs in the telophase when the major coils are being lost (figs. 5a, 19a, 65a).

There are three types of relational twisting of chromosomes: (1) twisting of non-homologous chromosomes; (2) twisting of paired chromosomes; (3) twisting of portions of the same arm or of the two arms of a chromosome. There is no need whatever to confuse any of these with relational coiling. First, because they are of chromosomes and relational coiling is always either of chromatids or half chromatids. Second, there are never more than a few gyres of the chromosomal twisting, while over three hundred gyres of relational coiling of the chromatids of a single chomosome may be seen plainly soon after they are produced (fig. 12b), and the same is true with regard to the number of gyres of relational coiling of the half chromatids of a single chromatid (fig. 62c).

If relational coiling is produced by torsion, as so many cytologists have speculated, why is it always confined to sister chromatids or to sister half chromatids?

Finally, in double chromatid chromosomes, there is unquestionable proof that relational coiling cannot be produced by one chromatid twisting about another. In these chromatids, from early prophase to metaphase (figs. 62–71), there are two generations of relational coiling, namely that of chromatids and that of half chromatids, and during the rest of the chromosome cycle

there is one generation. How, for example, could the twisting of one chromatid about another produce at the same time a relational coiling of the chromatids and a relational coiling of the half chromatids? The truth of the matter is: the relational coiling of the chromatids is present in the telophase (figs. 58b–60b) and it persists through telophase to prophase (figs. 61b–62b), without losing any gyres, and is present when the duplication of the chromatids produces relational coiling of half chromatids (fig. 62c). From this point on to metaphase, there is unwinding of the two generations of relational coiling, that of chromatids becoming completely unwound by metaphase, but that of half chromatids, having more unwinding to accomplish (and it cannot proceed faster than that of the chromatids), does not complete its unwinding. This relational coiling of half chromatids persists until the prophase of the next generation, when it becomes relational coiling of chromatids. Thus the two generations of relational coiling overlap one another. Is there any conceivable way such an overlapping could be accomplished by twisting?

In that type of prophase chromosome which is composed of two chromatids each of which is double, there must be four genes of each type, two of which are passed on to each gamete. A zygote formed by the fusion of such gametes has four, not two, genes of each type. On the other hand, when a prophase chromosome is composed of two single chromatids, only one gene of each type is passed on to each gamete, and the zygote formed by the fusion of these gametes has only two genes of each type. In the protozoa that I have studied, single chromatids are much more common than double ones, but in the higher forms of life double chromatids are probably more common. Can the same species in higher forms, as in *Holomastigotoides*, have both types of chromatids? What bearing do these cytological facts have on heredity? Do genes still behave as a unit irrespective of how many times they are represented? These questions cannot be answered at present. Huskins (1947) has discussed some of them in considerable detail.

SUMMARY

All chromosomes have minor and major coils; some have super coils; and some have supers in meiosis but not in mitosis. They all have a chromonema which, during a part of its life cycle, is embedded in a matrix; duplication can first be seen in the very early prophase when only minor coils are present and when chromosomes appear to be longest, although they are actually no longer in one stage than in another so far as their chromonemata are concerned. The minor coils are the only ones that exist throughout the life cycle of a chromosome; major and supers (when present) disappear and reappear during each chromosomal generation. Each of these three types of coils is independent in that minors do not grow up to become majors, nor do majors increase in size and become supers. Instead, minors are incorporated in majors and majors, together with the minors incorporated in them, are incorporated in supers, all of which is brought about by continuous contraction of the matrix, thus forcing the chromonema to occupy a shorter and wider space as contraction continues (figs. 46–51, 86). When majors and supers disappear, owing to the collapse and distintegration of the matrix, this process is reversed; the supers are lost first and then the majors. A chromosome goes through this cycle during each generation, that is, puts in major (and sometimes super) coils preparatory to nuclear and cell division and then loses these coils preparatory to duplication and separation (unwinding). But nuclear and cell division are in no way responsible for these changes, since they occur in endomitosis, a process where neither nucleus nor cell divides. Rather, these changes make it possible for a nucleus to divide when the proper stimuli are present. If a chromosome did not become apparently shorter, its daughters would not be free to move to separate poles; and, on the other hand, if it did not become apparently longer, by losing its major coils (and supers too when present), the relational coiling, produced as a result of duplication, could not be resolved and the daughters would remain coiled about each other.

Chromosomes are of two types, those with single and those with double chromatids.

Relational coiling occurs in chromatids and half chromatids, but not in chromosomes. Relational twisting occurs only in chromosomes and is of three types: paired homologous chromosomes, non-homologous chromosomes, and portions of the same arm or of the two arms of a chromosome. The twisting, as the term implies, is produced by two chromosomes or two parts of a single chromosome twisting about one another due to rotation. Relational coiling is not produced by any type of movement of the chromatids; no torsion or twisting is involved. It is produced by a duplication of the chromosome (or chromatid) in the same direction as that of the gyres of its ever present minor coils. Relational coiling is resolved by unwinding; each chromatid or half chromatid rotates.

"Individual" when applied to coiling is a better term than "internal." This is especially true since there is no such thing as external coiling. One, then, can speak of the individual minor, major, and super coils of chromosomes, chromatids, and half chromatids as contrasted with the relational coiling of chromatids and half chromatids.

There is no need for the retention of the term "standard coil" in mitosis.

Relic coiling is a misnomer; the major coils, of which it is supposed to be a remnant, disappear entirely in the very late telophase. What is left is merely waviness, twisting, and kinking of the chromonema, and this bears

TABLE 1

THE LIFE CYCLE OF CHROMOSOMAL TWISTING AND EACH TYPE OF COILING IN MITOSIS

Single Chromatids

	Prophase (resting)	Metaphase	Anaphase	Telophase
Minor coils	———			
Major coils	—————————————————————————————————————			
Super coils	—————————————————————————————————			
Relational coiling	————————			
Chromosomal twisting	———————————			————

Double Chromatids

	Prophase (resting)	Metaphase	Anaphase	Telophase
Minor coils	———			
Major coils	———————————————————————————————			
Super coils	—————————————————————————			
Relational coiling (two generations====) (one generation——)	=======			
Chromosomal twisting	———————			————

In meiosis I chromosomes, relational coiling and major coils are retained longer than in mitosis.

TABLE 2

THE SAME TYPE OF DIAGRAM AS SHOWN IN TABLE 1 EXCEPT THAT THE STARTING POINT IS IN THE TELOPHASE,
AS IN THE ILLUSTRATIONS, SHOWS SOME THINGS MORE CLEARLY

Single Chromatids

	Telophase	Prophase (resting)	Metaphase	Anaphase
Minor coils				
Major coils				
Super coils				
Relational coiling				
Chromosomal twisting				

Double Chromatids

	Telophase	Prophase (resting)	Metaphase	Anaphase
Minor coils				
Major coils				
Super coils				
Relational coiling (two generations====) (one generation——)				
Chromosomal twisting				

In meiosis I chromosomes, relational coiling and major coils are retained longer than in mitosis.

no relationship either to past or to future coils. Hence, twisting is a less confusing and more appropriate term than relic coiling.

The so-called resting stage in chromosomes is not the same in all types of cells; some chromosomes rest in the fairly late prophase, some in the telophase, and others in the interphase. Interphase and resting stages are not always synonymous. Failure to recognize this fact has produced much misunderstanding and confusion.

The centromere is certainly not merely a part of or a point on a chromosome. It leads an independent existence, and plays a more important role in mitosis and meiosis than previously recognized. In addition to its function in the movement of daughter chromosomes to the poles in mitosis and meiosis, it plays an important role in ·the reduction of chromosomes at meiosis.

There is no longer any question regarding the existence of terminal centromeres.

REFERENCES

CLEVELAND, L. R. 1938. Longitudinal and transverse division in two closely related flagellates. *Biol. Bull.* **74**: 1–24.

——1947a. Sex produced in the protozoa of Cryptocercus by molting. *Science* **105**: 16.

——1947b. The origin and evolution of meiosis. *Science* **105**: 287.

CLEVELAND, L. R., S. R. HALL, E. P. SANDERS, and J. COLLIER. 1934. The wood-feeding roach Cryptocercus, its protozoa, and the symbiosis between protozoa and roach. *Mem. Amer. Acad. Arts & Sci.* **17**: 185–342.

DARLINGTON, C. D. 1937. Recent advances in cytology. Ed. 2. London, Churchill.

DE MELLO, F. 1942. Hypermastiginids of the genus Holomastigotoides Grassi and Foa in the intestine of Hodotermes viarum Koenig from Coimbatore. *Arq. Escola med. cir. Nova-Goa Ser. A* **18**: 1–17.

GEITLER, L. 1938. Chromosomenbau. Berlin, Borntraeger.

GRASSI, B. 1917. Flagellati viventi nei Termiti. *Mem. R. Accad. Lincei* (5) **12**: 331–394.

GRASSI B., and A. FOA. 1911. Flagellati ai Protozoi dei Termitidi. Nota preliminare. *Rend. R. Accad. Lincei (Ser. 5)* **20**: 125–141.

HUGHES-SCHRADER, S., and H. RIS. 1941. The diffuse spindle attachment of coccids verified by the mitotic behavior of induced chromosome fragments. *Jour. Exp. Zool.* **87**: 429–456.

HUSKINS, C. L. 1947. The subdivision of the chromosomes and their multiplication in non-dividing tissues: possible interpretations in terms of gene structure and gene action. *Amer. Nat.* **81**: 401–434.

KOIDZUMI, M. 1921. Studies on the intestinal protozoa found in the termites of Japan. *Parasitology* **13**: 235–309.

MACKINNON, D. L. 1926. Observations on trichonymphids. I. The nucleus and axostyle of Holomastigotoides hemigymnum Grassi (?). *Quart. Jour. Micro. Sci.* **70**: 173–191.

NOBLE, E. R. 1947. Mitosis in Entamoeba gingivalis. *Anat. Rec.* **99**: 47.

RAFALKO, J. S. 1947. Mitotic division in the amoebo-flagellate, Tetramitus rostratus. *Anat. Rec.* **99**: 48.

SCHAEDE, R. 1936. Untersuchungen mit des Nuklealreaktion an Kern und Kernteilung. *Planta* **26**: 167–192.

SCHRADER, R. 1935. Some notes on the behavior of long chromosomes. *Cytologia, Tokyo* **6**: 622–630.

——1936. The kinetochore or spindlefiber locus in Amphiuma tridactylum. *Biol. Bull.* **70**: 484–498.

——1939. The structure of the kinetochore at mitosis. *Chromosoma* **1**: 230–237.

SPARROW, A. H. 1942. The structure and development of the chromosome spirals in microspores of Trillium. *Can. Jour. Res. C* **20**: 257–266.

WHITE, M. J. D. 1945. Animal cytology and evolution. Cambridge Univ. Press.

EXPLANATION OF PLATES

Unless otherwise noted, drawings are at a magnification of 2100, fixative is Schaudinn's, stain is iron alum hematoxylin, organism is *Holomastigotoides tusitala,* and host is *Prorhinotermes simplex* Hagen. In *H. tusitala* and *H. diversa* long chromosomes, chromatids, and half chromatids, with lateral and terminal nucleoli, are dark; short chromosomes, chromatids, and half chromatids, with only terminal nucleoli, are light. Each nucleus of *H. tusitala,* no matter how many times it is drawn, is given the same number, but each drawing is given a different letter, and each letter, throughout the plates, refers to the same type of coiling:

a—twisting of chromosomes (individual, relational, or both)
b—relational coiling of chromatids
c—relational coiling of half chromatids
d—major coils of chromosomes, chromatids, half chromatids
e—minor coils of chromosomes, chromatids, half chromatids.

27

PLATE 1

FIG. 1a. Telophase shortly after cytoplasmic division. Anterior daughter cell. Flagellar bands viewed somewhat vertically showing the manner in which they arise in anterior end of cell. New, fifth band and new centriole that follows it have grown out to point where line is broken. Early individual twisting of chromosomes. Both chromosomes are anchored by remnants of the achromatic figure to old centriole which leaves flagellar band 5 after following it for 1½ turns.

FIG. 1d. Major coils. Reversals of majors are of no significance when method of their disappearance and reappearance in each generation of chromosomes is clearly understood.

FIG. 1e. Minor coils incorporated in majors. Chromosomal membranes are fusing to form a new, common nuclear membrane.

FIG. 2a. Chromosomes appear longer and there is only a small amount of individual twisting in them. New flagellar band 5 and centriole have made further progress in growing out and taking their places beside old band 4 and the old centriole that follows it. Note how new centriole is leaving flagellar band 5 at the same point where the old one leaves band 4.

FIG. 2d. Major coils and more of them than in 1a.. The number varies considerably from cell to cell.

FIG. 2e. Minor coils.

FIG. 3a. Considerable individual twisting in each chromosome. Nuclear membrane becomes shorter and wider as chromosomes twist. New flagellar band 5 and new centriole that follows it have completed their growth, the new and old centrioles have formed an achromatic figure between their free distal ends, one chromosome (the short one in this cell) has moved from its former anchorage to the old centriole and is now fastened to the new centriole (lower right corner of the achromatic figure). The long chromosome has not moved, has kept its anchorage to the old centriole (upper left corner of achromatic figure). See text-figures 2 and 3.

FIG. 3d. Major coils are loose and in places have lost some of their regularity.

FIG. 3e. Minor coils are placed under slight tension in those places where majors are losing their regularity.

28

PLATE 1

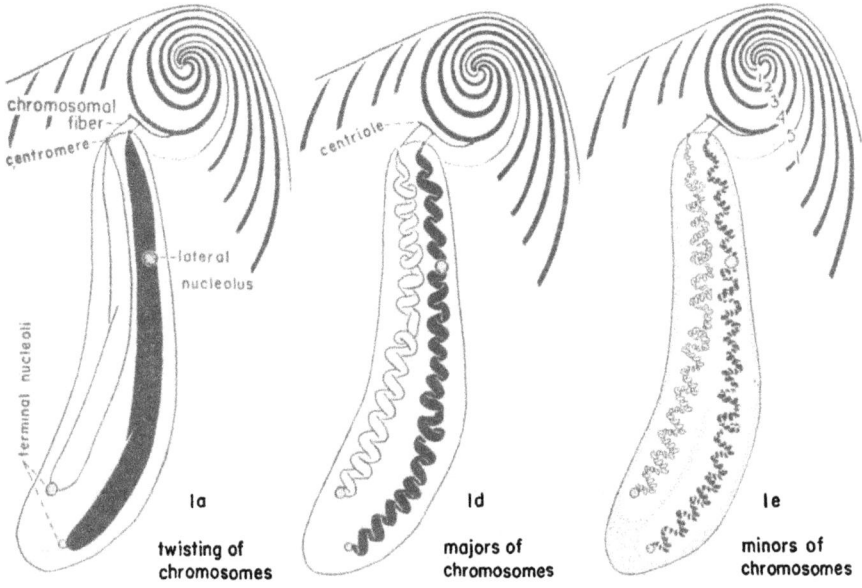

1a
twisting of
chromosomes

1d
majors of
chromosomes

1e
minors of
chromosomes

TELOPHASES

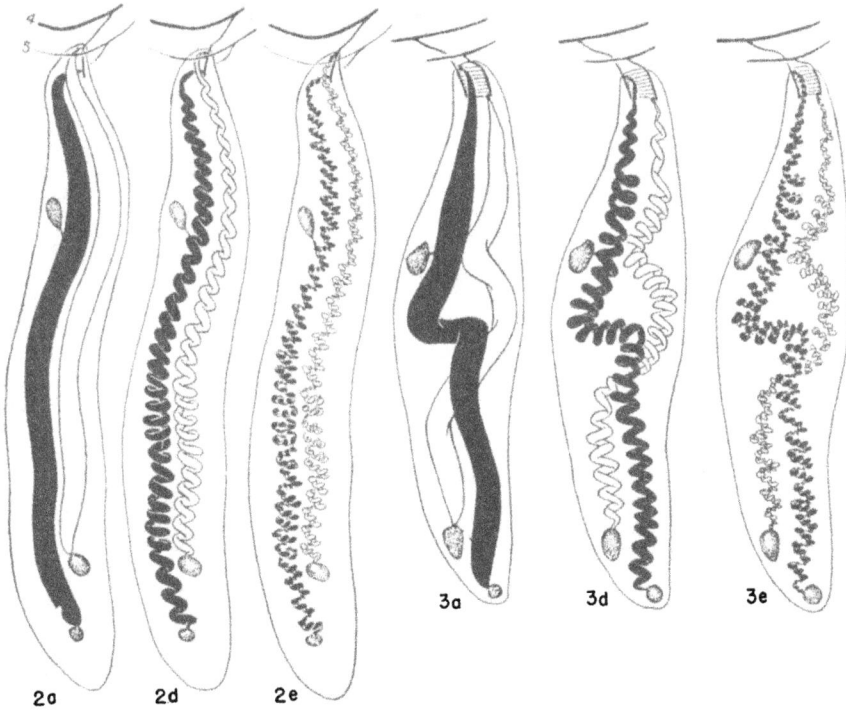

2a

2d

2e

3a

3d

3e

PLATE 2

Fig. 4a. More individual twisting of chromosomes, nucleus shorter and wider. The flagellar bands (4 and 5) slant differently from those of figs. 2a and 3a because they are the bottom bands. Those of 2a and 3a are the top bands. See text-figure 1, b. In later stages the free ends of the centrioles usually extend from the bands at right angles or nearly so and the achromatic figure is more or less parallel with the long axis of the nucleus.

Fig. 4d. Major coils looser and chromosomes appear longer.

Fig. 4e. Minors are almost free near centromere ends because majors at these places have nearly disappeared.

Fig. 5a. Slightly more individual twisting of chromosomes and a gyre of relational twisting.

Fig. 5d. Majors more irregular, not tightly coiled in any portion of either chromosome. By this stage the collapsing matrix material has begun to scatter through the nucleus.

Fig. 5e. Note one portion of short (light) chromosome near centromere end where only minors are present.

Fig. 6a. Individual twisting of chromosomes more advanced. One gyre of relational twisting. Even though greatly twisted, chromosomes are longer than nucleus and are bending on themselves anteriorly.

Fig. 6d. Major coils very irregular for most of the length of each chromosome. Note how the chromonema loops on itself as the majors are released by the collapsing matrix.

Fig. 6e. Only minors present in many places along each chomosome.

PLATE 2

twisting of
chromosomes

4a

majors of
chromosomes

4d

minors of
chromosomes

4e

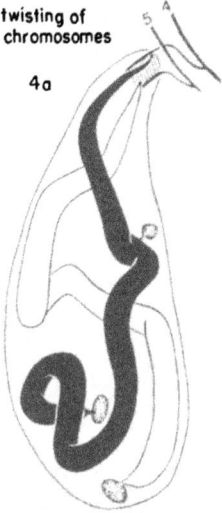
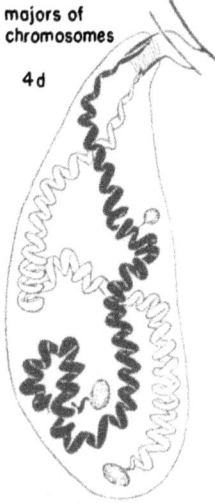
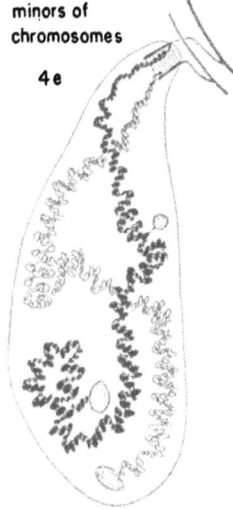

5a

5d

5e

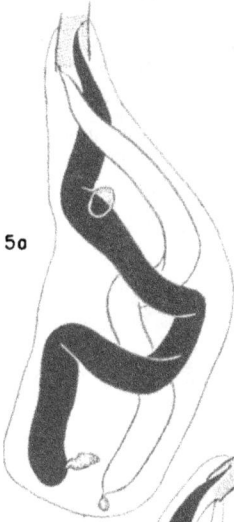
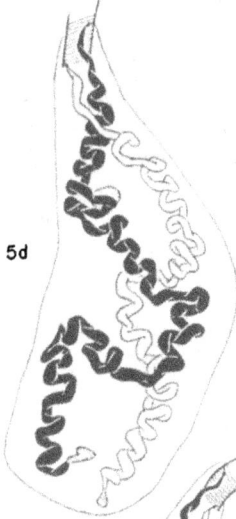
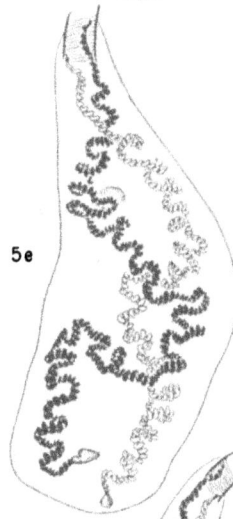

T
E
L
O
P
H
A
S
E
S

6a

6d

6e

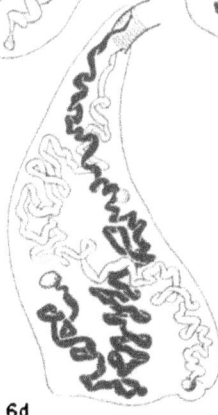
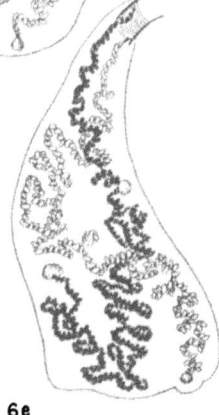

PLATE 3

FIG. 7a. More individual twisting of chromosomes but no relational twisting. Terminal nucleoli have fused.

FIG. 7d. Chromonema folds back on itself more and thus occupies more of the nucleus which has become broader—an adjustment to accommodate the apparent lengthening of the chromosomes. In only a few places can one see more than remnants of major coils. In the region between nucleoli not even a remnant of them is left.

FIG. 7e. Even though major coils are nearly gone minors are very tight all along each chromosome.

FIG. 8a. Slightly more individual twisting of chromosomes. All nucleoli—two terminals and lateral—fused.

FIG. 8d. Still more regions along each chromosome where no major coils at all are present; in a few places remnants of them can be seen.

FIG. 8e. Minor coils are not as tight in those places (lower portion of nucleus) where majors have disappeared completely as where remnants remain (near nucleoli and achromatic figure).

FIG. 9a. Chromosomes greatly twisted individually but only one gyre of relational twisting. No fusing of nucleoli.

FIG. 9d. Remnants of major coils near centromere ends of each chromosome and a short portion of the long chromosome in centre of nucleus. Elsewhere along each chromosome not a vestige of major coils remains.

FIG. 9e. Minors looser except in those places where remnants of the major coils remain. Nucleus is broader and the two chromonemata fill most of it.

PLATE 3

TELOPHASES

twisting of chromosomes   majors of chromosomes   minors of chromosomes

7a

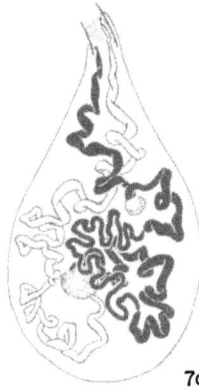

7d

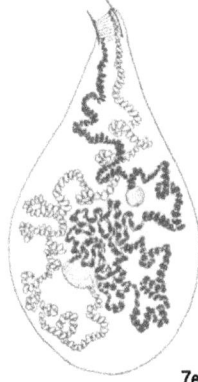

7e

8a

8d

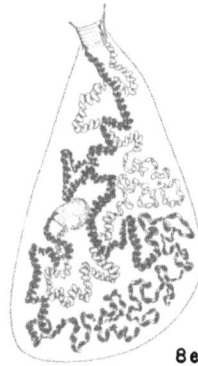

8e

last remnant of major coils

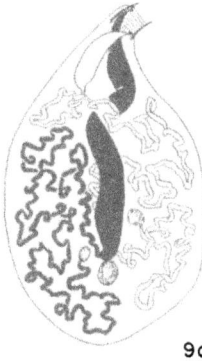

9a

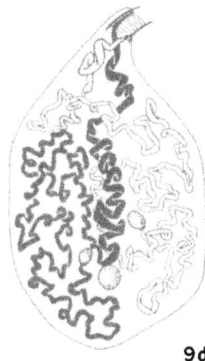

9d

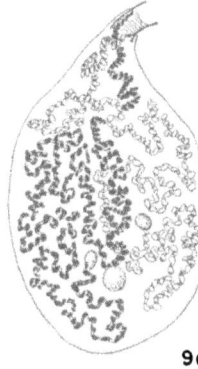

9e

PLATE 4

Fig. 10g. Chromosomes greatly twisted individually, but not relationally. All nucleoli fused.

Fig. 10e. Matrix completely collapsed and hence exerts no tension on the minor coils, allowing them to relax and thus reveal their true molecular configuration. This is the so-called "interphase" of many types of cells and it is also the resting stage for many cells, but not for all, not for *Holomastigotoides.*

Fig. 11a. There is a great deal of individual twisting of chromosomes, but less than a gyre of relational twisting (the short chromosome lies over the long one). × 3,100.

Fig. 11e. There are places in each chromosome where duplication has not occurred and only minor coils are present; other places where duplication has occurred and has produced tight relational coiling of chromatids, each of which has its own minor coils; and a few places (*z* and near centromeres) where duplication has been followed by some unwinding of the relational coiling. Centromeres were duplicated at the same time as the chromosomes, and one is connected to each corner of the achromatic figure, but only two of them—one connected to a short and one to a long chromatid—have moved. × 3,100.

34

PLATE 4

TELOPHASE

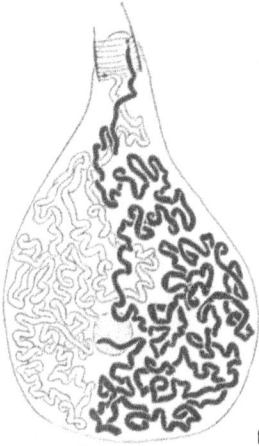

10a

twisting of chromosomes

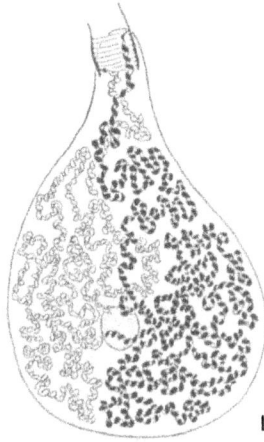

10e

minors of chromosomes

PROPHASE

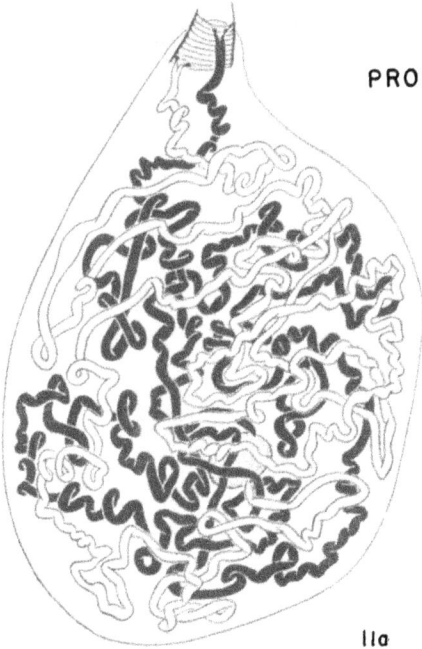

11a

twisting of chromosomes

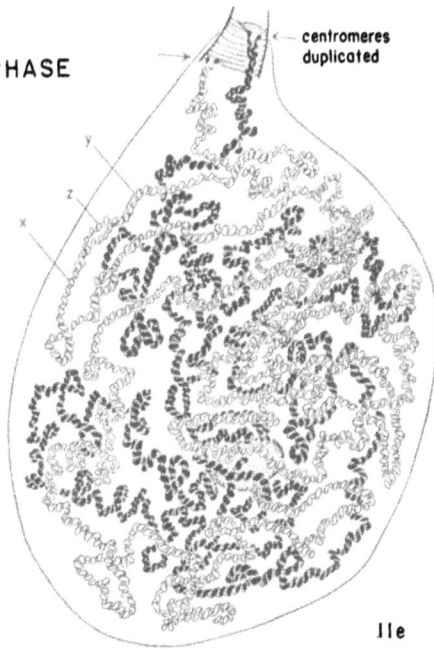

centromeres
duplicated

y

z

x

11e

minors of chromosomes and chromatids
chromonemata duplicating
x—not duplicated
y—duplicated, not unwound
z—duplicated, unwound

PLATE 5

FIG. 12a. Both individual and relational twisting of chromosomes. Terminal nucleoli fused.

FIG. 12b. Relational coiling of chromatids is evident all along each chromosome, being still fairly tight in most places. Unwinding does not have to begin at ends of relationally coiled chromatids. It may begin anywhere.

FIG. 12e. In those places where relational coiling of chromatids is unwound, minor coils of each chromatid are plainer.

FIG. 13a. Individual and relational twisting of chromosomes.

FIG. 13b. Considerable progress in unwinding of relationally coiled chromatids.

FIG. 13e. Minor coils may now be seen plainly all along each chromatid. There is no difficulty in differentiating relational and minor coils as in earlier stages with less unwinding of relationally coiled chromatids.

FIG. 14a. Individual twisting of chromosomes. Lateral nucleoli of long chromatids fused with terminal nucleoli of short chromatids.

FIG. 14b. Relational coiling of chromatids out in places; varying degrees of tightness elsewhere.

FIG. 14e. Contraction of the matrix of each chromatid has tightened minor coils considerably.

36

PLATE 5

twisting of
chromosomes

12 a

↓

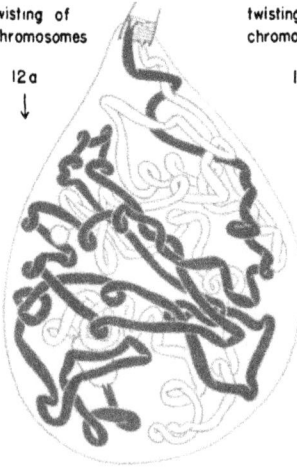

twisting of
chromosomes

13 a

↓

PROPHASES

twisting of
chromosomes

14 a

↓

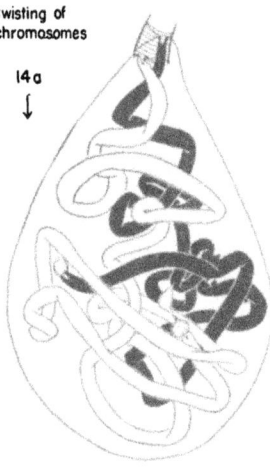

rela. coiling of
chromatids

12 b

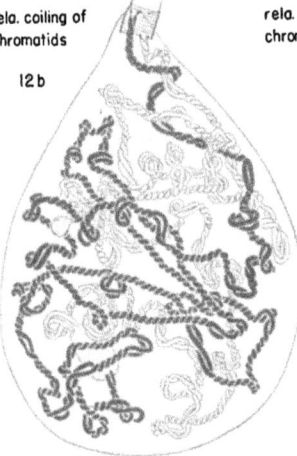

rela. coiling of
chromatids

13 b

rela. coiling of
chromatids

14 b

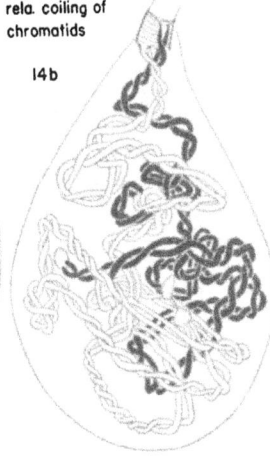

minors of
chromatids

12 e

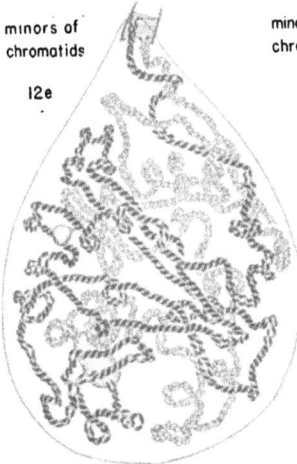

minors of
chromatids

13 e

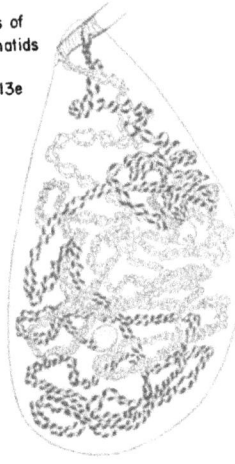

minors of
chromatids

14 e

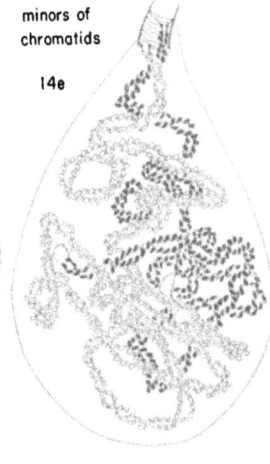

PLATE 6

FIG. 15a. Individual twisting of chromosomes is becoming less. One gyre of relational twisting of chromosomes.

FIG. 15b. Relational coiling of chromatids is out for longer stretches. However, there is still a fair amount left.

FIG. 15e. Minor coils are as tight—close together—as they can be. More contraction of matrix will force chromonema to bend, throw minors out of alignment, and thus start the incorporation of minors in a new generation of major coils.

FIG. 16a. Less individual twisting of chromosomes.

FIG. 16b. Not many gyres of relational coiling of chromatids left.

FIG. 16d. The bends in the chromonemata of the chromatids are the beginning of a new generation of major coils of chromatids.

FIG. 16e. Contraction of the matrix of each chromatid could not force the minor coils of the chromonema any closer together, so it pushed them out of line, bent the chromonema, and in so doing began to incorporate the minors in developing new, major coils.

PLATE 6

PROPHASES

twisting of
chromosomes

15a

rela. coiling of
chromatids

15b

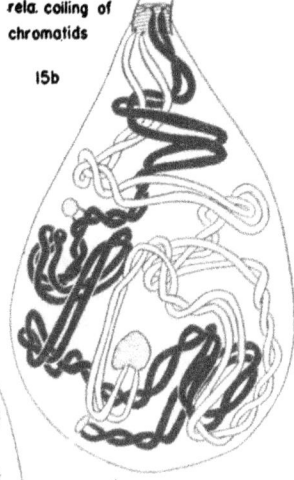

minors of chromatids

15e

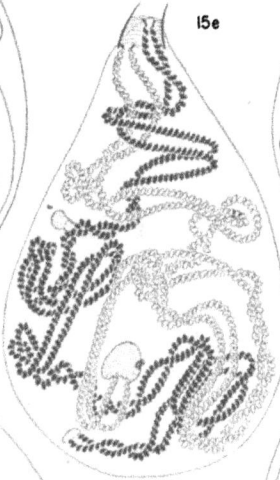

twisting of
chromosomes

16a

rela. coiling of
chromatids

16b

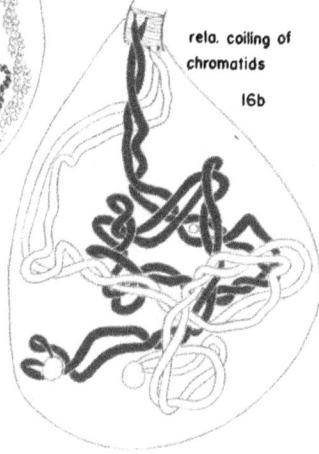

majors of
chromatids

16d

beginning of
major coils

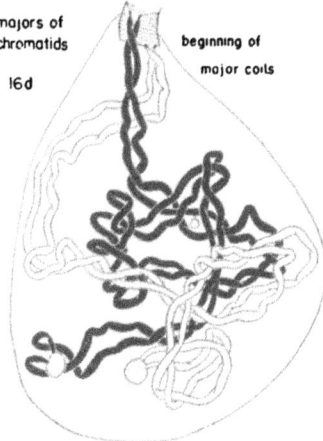

minors of
chromatids

16e

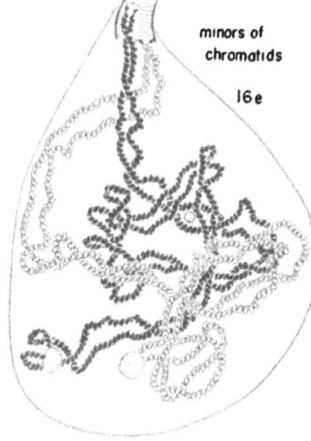

PLATE 7

FIG. 17a. The chromosomes are apparently shorter and have less individual twisting.

FIG. 17b. Only a few gyres of relational coiling of chromatids left. All nucleoli have been fused and the terminals are pulling away from the laterals.

FIG. 17d. In some places the bends of the chromonemata are becoming better organized into major coils. It is plain how more contraction of matrix will produce tighter and more regular major coils.

FIG. 17e. Minor coils are slowly being incorporated in the developing major coils.

FIG. 18a. Still less twisting of chromosomes.

FIG. 18b. Only three gyres of relational coiling in the long and only one in the short chromatids. Terminal nucleoli of all four chromatids fused.

FIG. 18d. Major coils, owing to continued contraction of the matrix, are becoming more and more regular. In a few places they are almost as tight as they will ever be.

FIG. 18e. Continued progress in the incorporation of minors in majors.

PLATE 7

twisting of
chromosomes

17a.

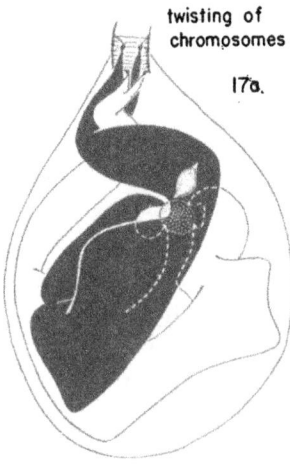

rela. coiling of
chromatids

17b

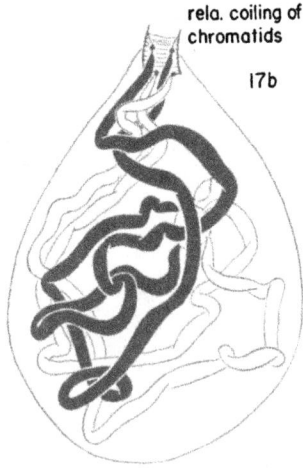

majors of
chromatids

17d

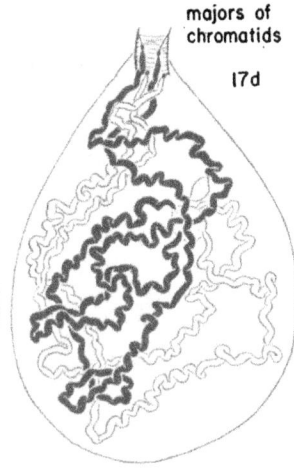

minors of
chromatids

17e

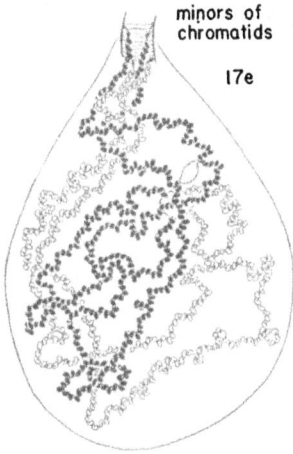

PROPHASES

twisting of
chromosomes

18a

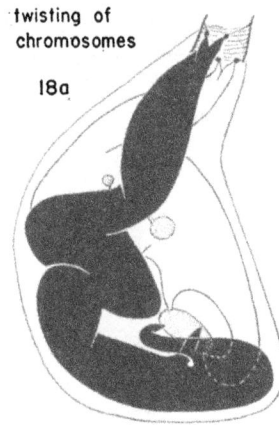

rela. coiling of
chromatids

18b

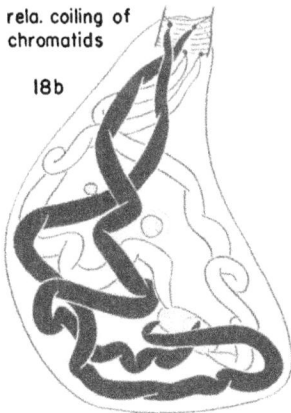

majors of
chromatids

18d

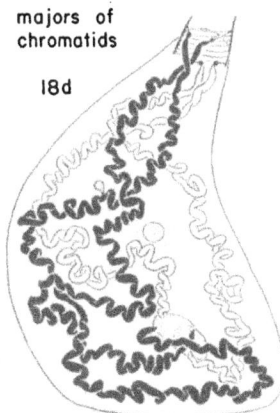

minors of
chromatids

18e

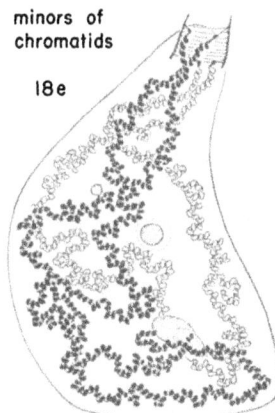

PLATE 8

FIG. 19a. Almost two gyres of relational twisting of chromosomes.

FIG. 19b. About two gyres of relational coiling of chromatids. It is easy to see that when this coiling comes out the chromosomal twisting will come out too. No fusion of nucleoli.

FIG. 19d. Considerably more regularity in the major coils all along each chromatid.

FIG. 19e. There are now no minor coils that are not incorporated in majors.

FIG. 20a. Still less individual twisting of chromosomes.

FIG. 20b. Now less than two gyres of relational coiling in chromatids. If terminal nucleoli from two long and one short chromatids were not fused, there would be only one gyre of relational coiling. Lateral nucleoli fused.

FIG. 20d. Major coils broader, more regular, thus making chromatids appear shorter.

FIG. 20e. Enlargement of majors incorporates more minors in them.

42

PLATE 8

twisting of
chromosomes

19a

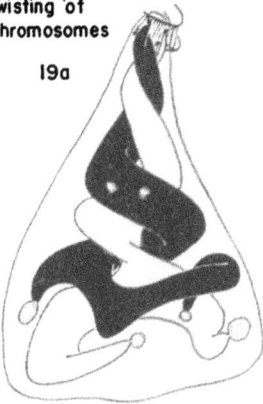

rela. coiling of
chromatids

19b

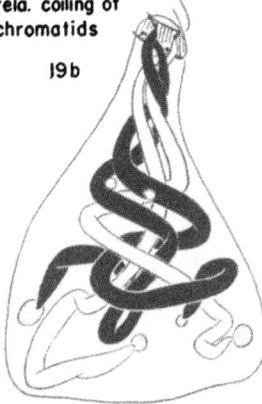

majors of
chromatids

19d

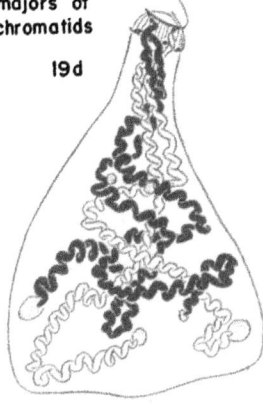

minors of
chromatids

19e

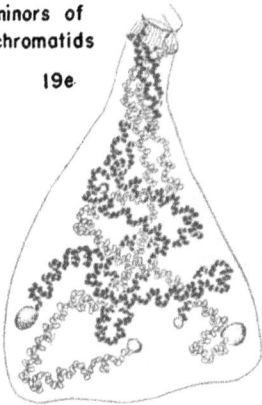

PROPHASES

twisting of
chromo-
somes

20a

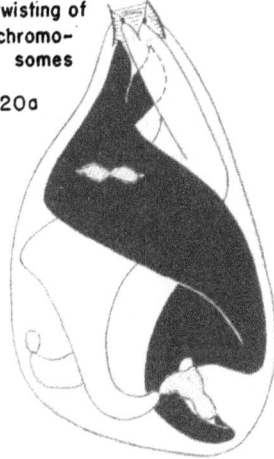

rela. coiling
of chrom-
atids

20b

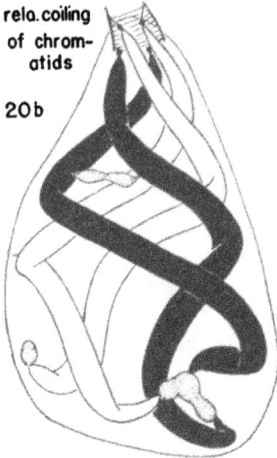

majors of
chroma-
tids

20d

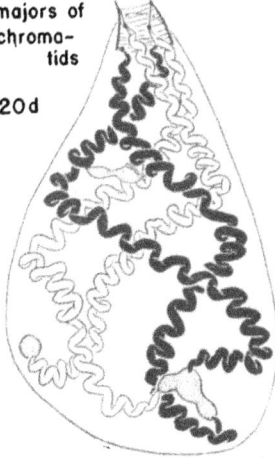

minors of
chroma-
tids

20e

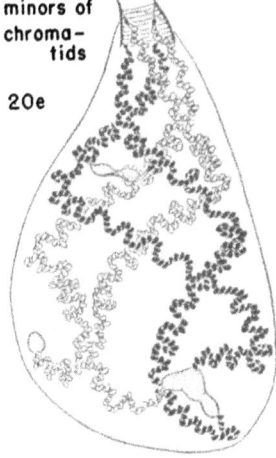

PLATE 9

FIG. 21a. Twisting of chromosomes has all but disappeared and will not be shown again until it reappears in late telophase.

FIG. 21b. Less than a gyre of relational coiling of chromatids remains. No nucleoli fused.

FIG. 21d. Majors closer together and therefore chromatids appear shorter.

FIG. 21e. Minors in majors very plain. Membranes forming around each chromatid. All nuclear material not within these membranes will be slowly discarded into the cytoplasm as development progresses. After cell division, the individual chromosomal membranes fuse to form the new nuclear membrane.

FIG. 22b. Relational coiling of chromatids is nearly all out and will not reappear in the life cycle of the chromosomes until the next early prophase.

FIG. 22d. Major coils as close together and regular as they will ever be. Fusion of terminal nucleoli of non-sister chromatids.

FIG. 22e. Except for membranes around each chromatid, this is the condition of most organisms in any preparation—and is the resting condition. Formation of membranes around chromatids is the first step in prophase development following the resting stage.

PLATE 9

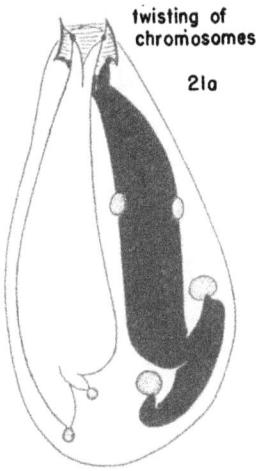

twisting of
chromosomes

21a

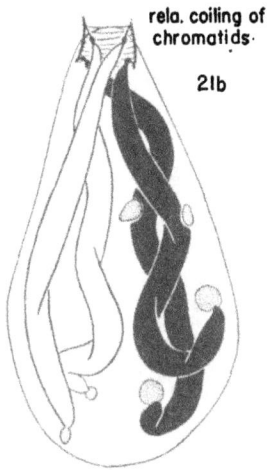

rela. coiling of
chromatids

21b

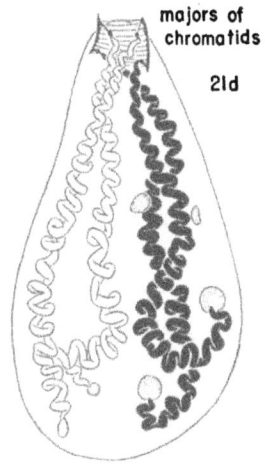

majors of
chromatids

21d

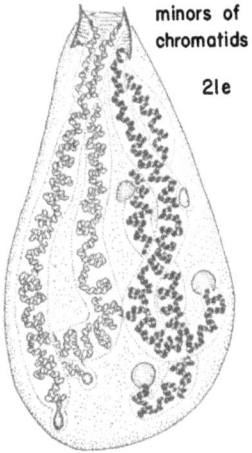

minors of
chromatids

21e

PROPHASES

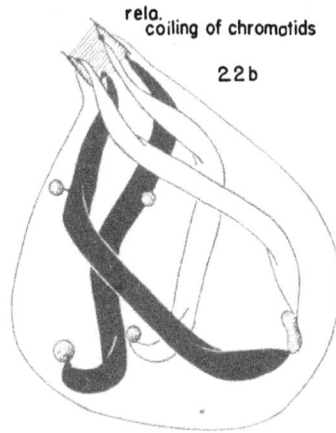

rela.
coiling of chromatids

22b

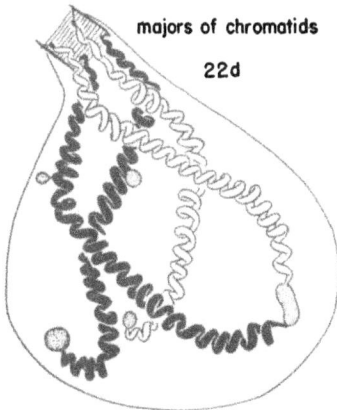

majors of chromatids

22d

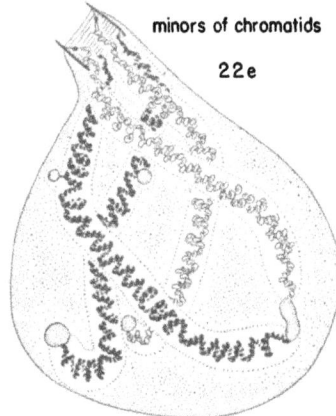

minors of chromatids

22e

PLATE 10

FIG. 23d. Prophase development following resting stage. Achromatic figure has begun to grow again, becoming longer and wider. As it increases in length, sister chromatids are slowly separated, separation at the centromere ends occurring before it does elsewhere.

FIG. 23e. Minor coils. Pieces of nuclear membrane, together with contents within nucleus, have been discarded into the cytoplasm.

FIG. 24d. Sister chromatids slightly more separated than in previous illustration.

FIG. 24e. Minor coils. More progress in discarding of pieces of nuclear membrane and nuclear contents.

PLATE 10

majors of
chromatids

23d

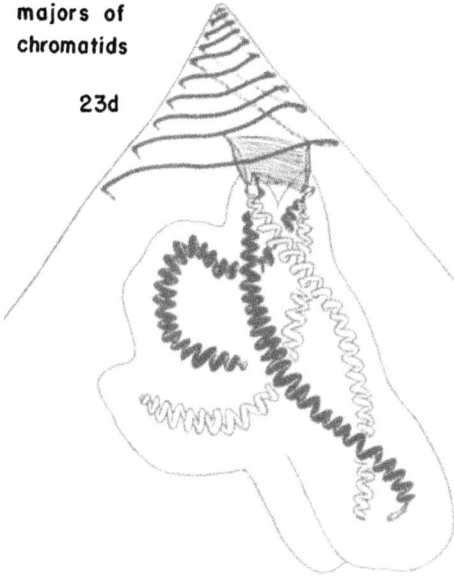

minors of
chromatids

23e

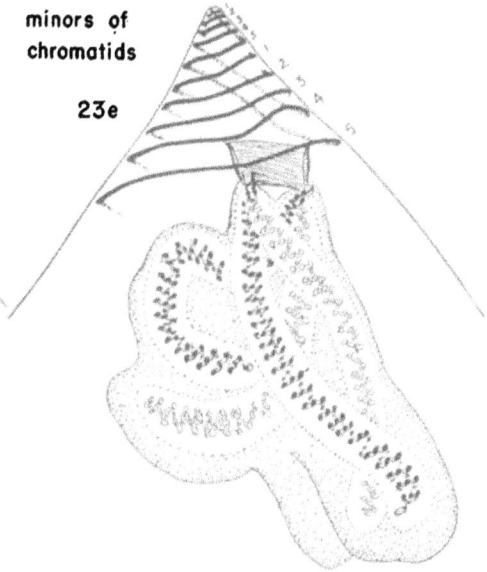

PROPHASES

majors of
chromatids
24d

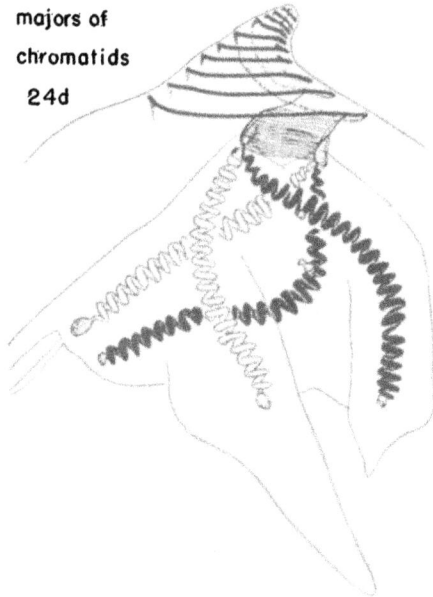

minors of
chromatids
24e

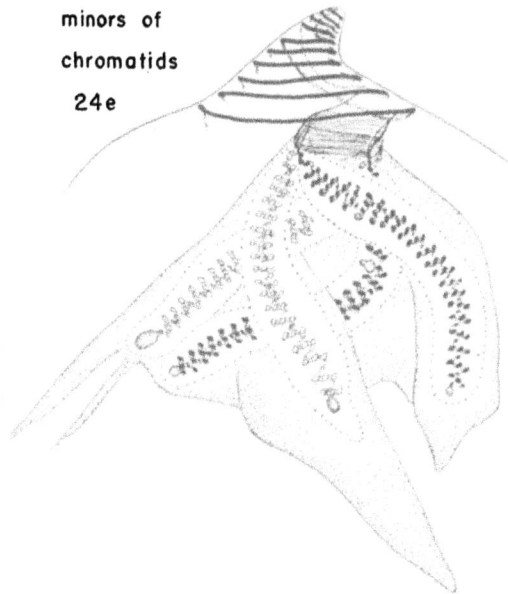

PLATE 11

FIG. *25d.* This is about the only stage one can label metaphase. There is no true metaphase stage. The centromere ends of sister chromatids are in anaphase. They have been separated since very early prophase. The distal halves of the chromatids have not separated at all.

FIG. *25e.* Minor coils. Anteriorly the nuclear membrane is dividing; much of it has yet to be discarded.

FIG. *26d.* Major coils. Central spindle has elongated. The centrioles and flagellar bands are more rigid than central spindle, so it was forced to bend as it grew. Thus the sister chromatids are separated and two non-sister chroma-tids come to lie side by side. This is the only time in the life cycle of the chromosomes of *H. tusitala* that any attempt at pairing of chromatids is seen. It is a very feeble one, occurring infrequently and mostly near distal ends. Non-sisters pair—if indeed it can be called pairing at all—as well as sisters. In the closely related *H. diversa,* typical pairing is a regular occurrence, the haploid variety forming dyads and the diploid tetrads.

FIG. *26e.* Minor coils. Nucleus is much longer than chromosomes. If anterior ends of sister chromatids were not so far apart, this could easily be called a metaphase.

PLATE 11

METAPHASE

majors of chromatids

25d

minors of chromatids

25e

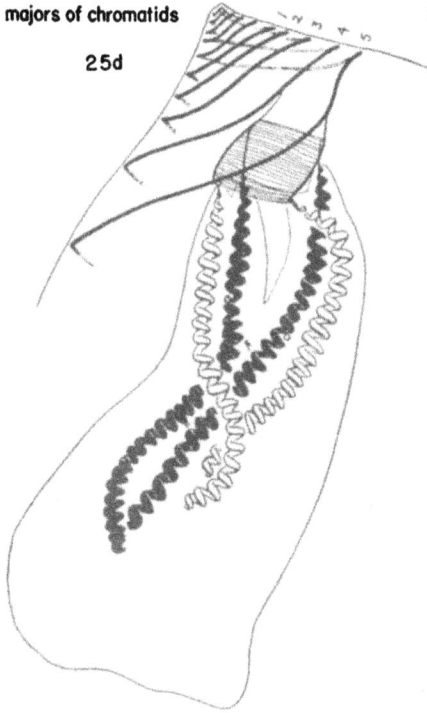
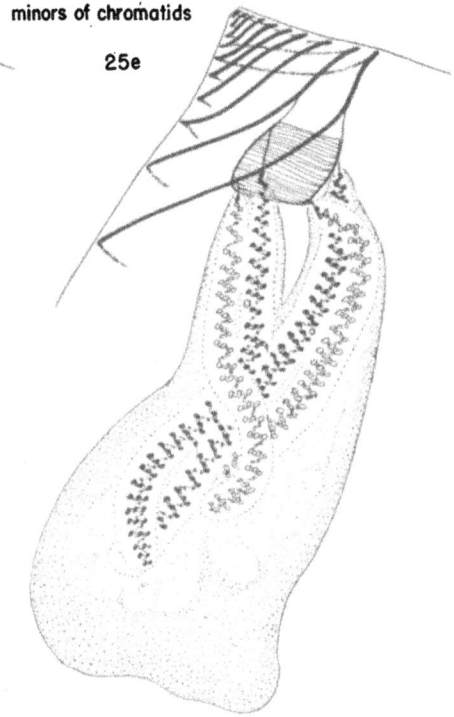

ANAPHASE

majors of chromatids

26d

minors of chromatids

26e

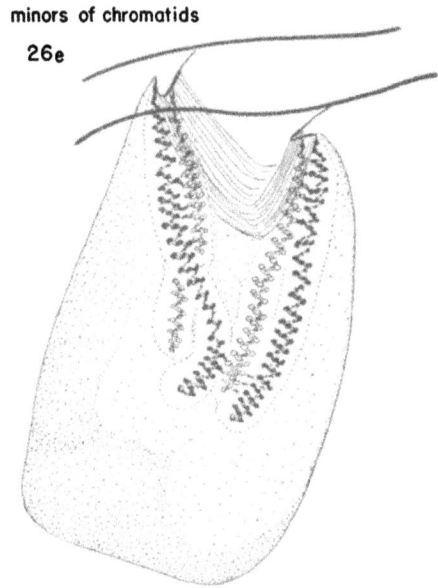

PLATE 12

FIG. *27d.* This organism is at about the same stage of development as the previous one, but appears different since it is viewed from a different angle. The two would be much alike if one were rotated 180°. The centrioles are leaving the flagellar bands from the bottom.

FIG. *27e.* Minor coils have not changed at all since late prophase.

PLATE 12

ANAPHASE

majors of chromatids

27d

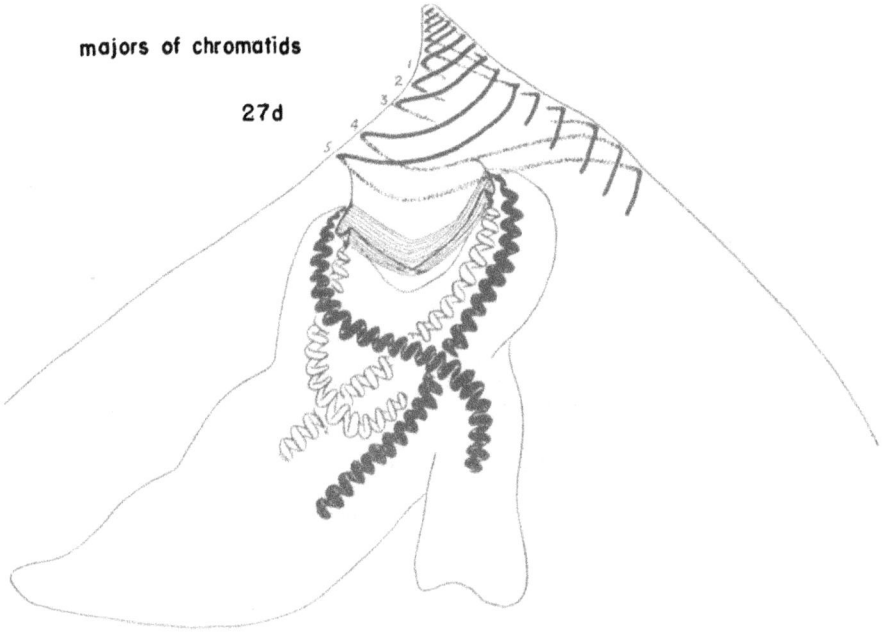

minors of chromatids

27e

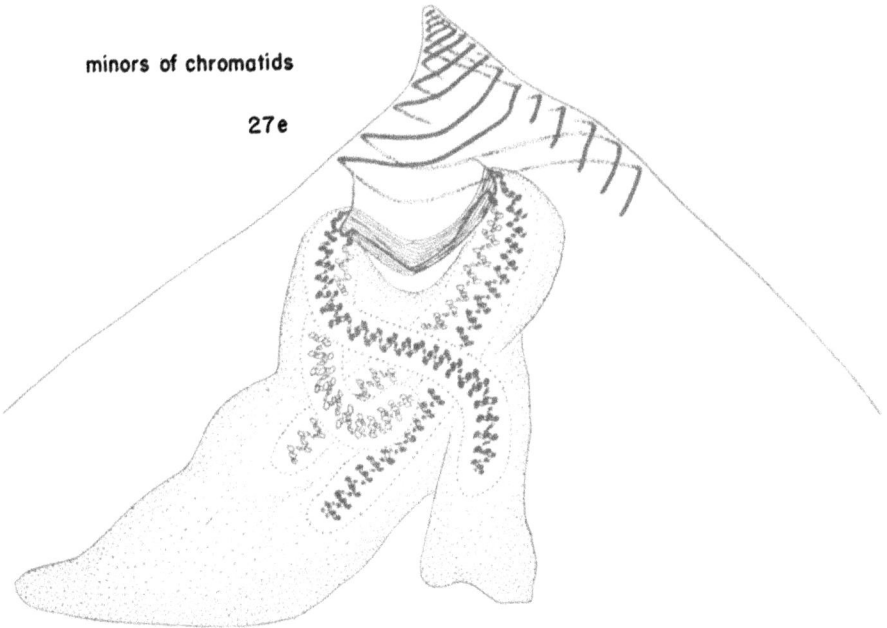

PLATE 13

FIG. *28d*. Increase in length of the fibres (astral rays) of the central spindle has bent it greatly, producing almost a circle. The central spindle could not push the flagellar bands and centrioles apart. The non-sister chromatids are in two well-separated groups and I suppose this can be called a telophase.

FIG. *28e*. The membranes around each chromatid are clearly demarked. Everything else in the nucleus will be discarded eventually.

FIG. *29d*. Growth of the central spindle fibres has not progressed faster than development and separation of chromatids, so the central spindle in this cell has not been forced to bend. Another example of homologous and non-homologous "pairing." Joining of last gyre of one long (dark) chromatid with 14th gyre (from distal end) of its sister has retarded separation.

FIG. *29e*. Some of the interconnections between chromatids clearly extend from the chromonema of one to that of the other.

PLATE 13

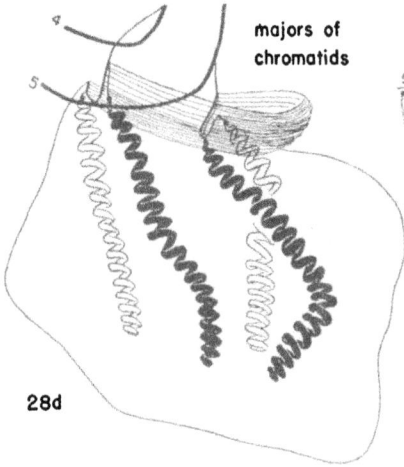

majors of
chromatids

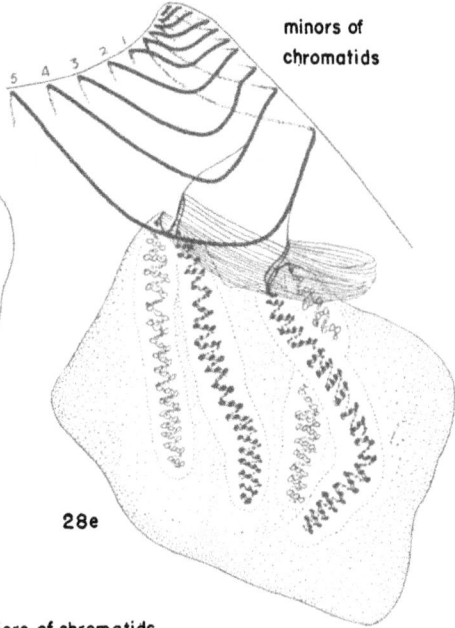

minors of
chromatids

28d

28e

TELOPHASES

majors of chromatids

29d

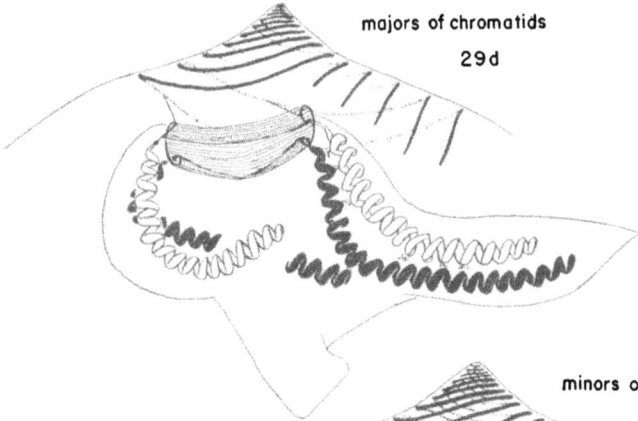

minors of chromatids

29e

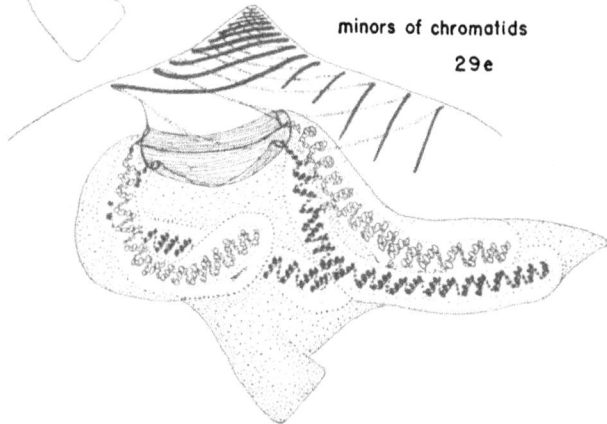

PLATE 14

Fɪɢ. 30d. Flagellar band 5 has separated from the other four
    bands anteriorly, and has made 1½ turns of the body of the
    cell. Its free, anterior end now lies adjacent to the nucleus.
    Before this band separated from the others, the central
    spindle had become greatly bent, and remains so. Now, the
    nucleus can divide since one flagellar band and its centriole
    will be free to move and carry two of the chromatids with
    it as soon as the central spindle pulls apart, and this occurs
    presently.
Fɪɢ. 30e. Nucleus lobulated, preparatory to discarding of large
    portions of its contents into cytoplasm.

PLATE 14

TELOPHASE

majors of chromatids

30d

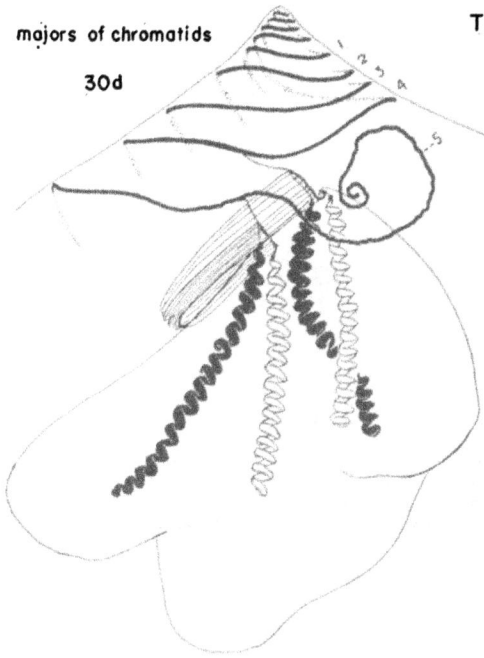

minors of chromatids

30e

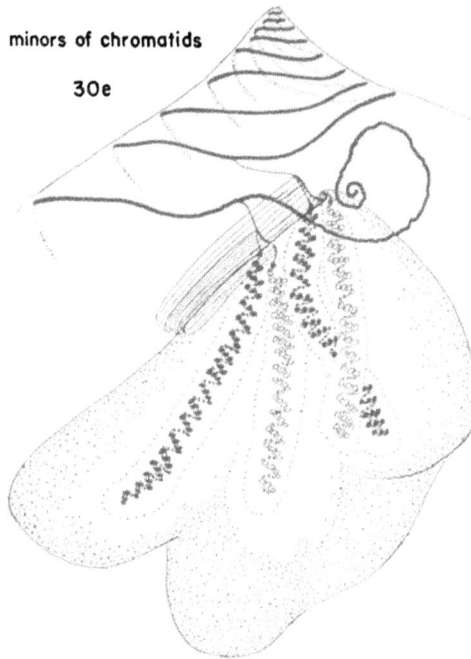

PLATE 15

Fig. 31*d*. Further progress in unwinding of flagellar band 5, central spindle has pulled apart, nucleus has discarded more of its contents and is dividing.

Fig. 31*e*. Minor coils are still incorporated in tight major coils.

PLATE 15

majors of chromatids

31d

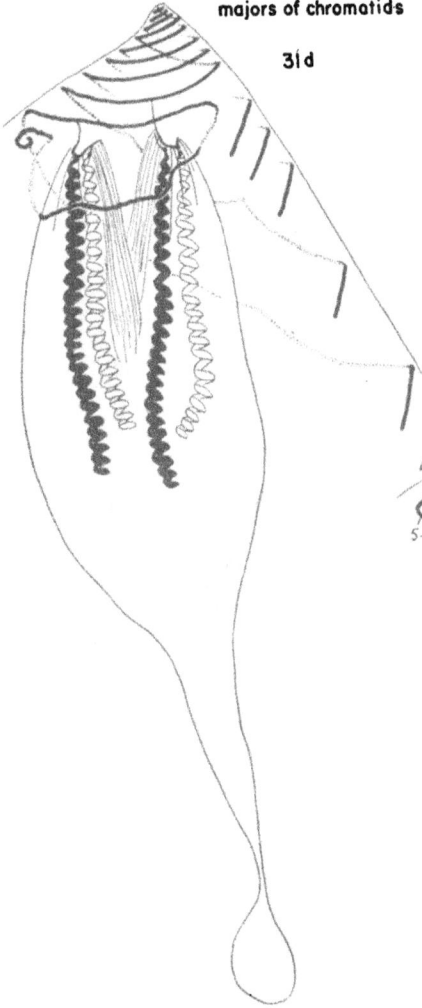

TELOPHASE

minors of chromatids

31e

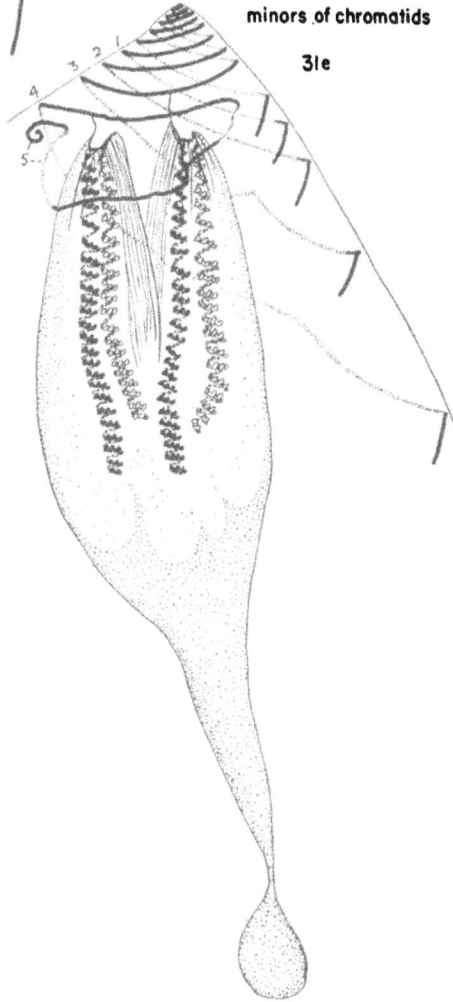

PLATE 16

Fig. *32d*. Nuclear division complete and unequal except for chromatids. The smaller daughter nucleus always goes with the free, fifth flagellar band to which it is connected by the centriole and the remains of that portion of the achromatic figure which this centriole produced. The other nucleus, in the same manner, is connected with the fourth flagellar band. This band does not move; hence the nucleus attached to it must remain stationary.

Fig. *32e*. Each nucleus still has more material to discard; the free one is farther along and will complete the progress sooner than the larger, stationary one. More progress has been made in unwinding of flagellar band 5.

PLATE 16

TELOPHASE

majors of chromatids

32d

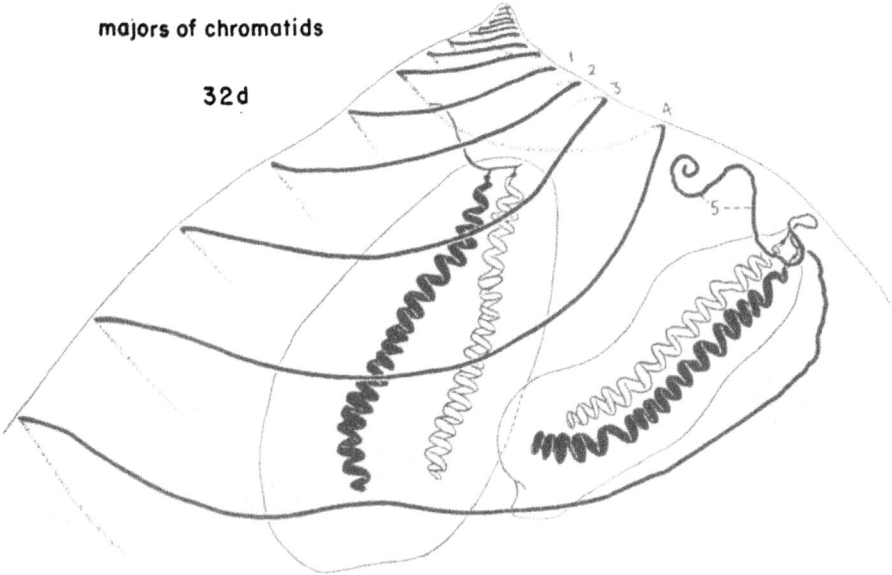

minors of chromatids

32e

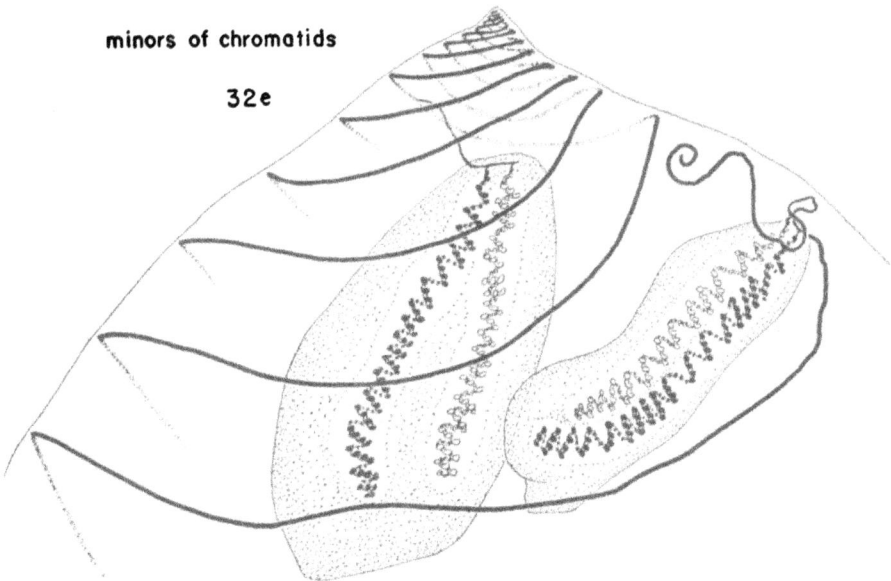

PLATE 17

FIG. *33d.* The free flagellar band has made more progress in unwinding and separating itself from the other four bands, and is carrying its nucleus with it. The major and minor coils do not change at all during the brief period from this stage to the telophase after cytoplasmic division (figs. 1*d*, 1*e*) where the account of them began. Since they do not change and since the primary interest from this point on to the completion of cell division is in the unwinding of flagel-lar band 5, only entire cells at lower magnifications with major and minor coils omitted are illustrated.

FIG. *33e.* The nucleus attached to the free, fifth flagellar band has now discarded all of its material except that which from the late prophase on has been contained within the small membrane around each chromatid. The other nucleus will soon do the same thing. It always does so later.

PLATE 17

TELOPHASE

majors of chromatids

33 d

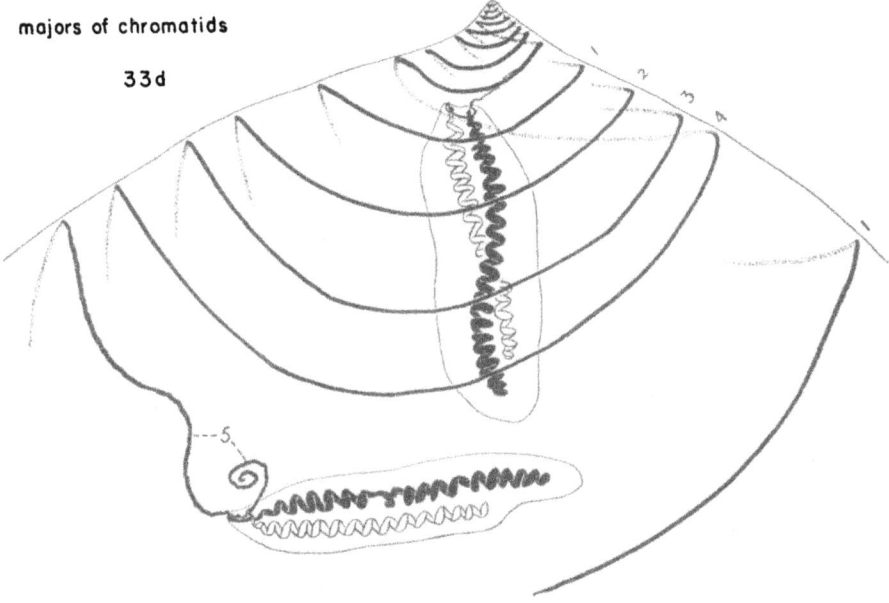

minors of chromatids

33 e

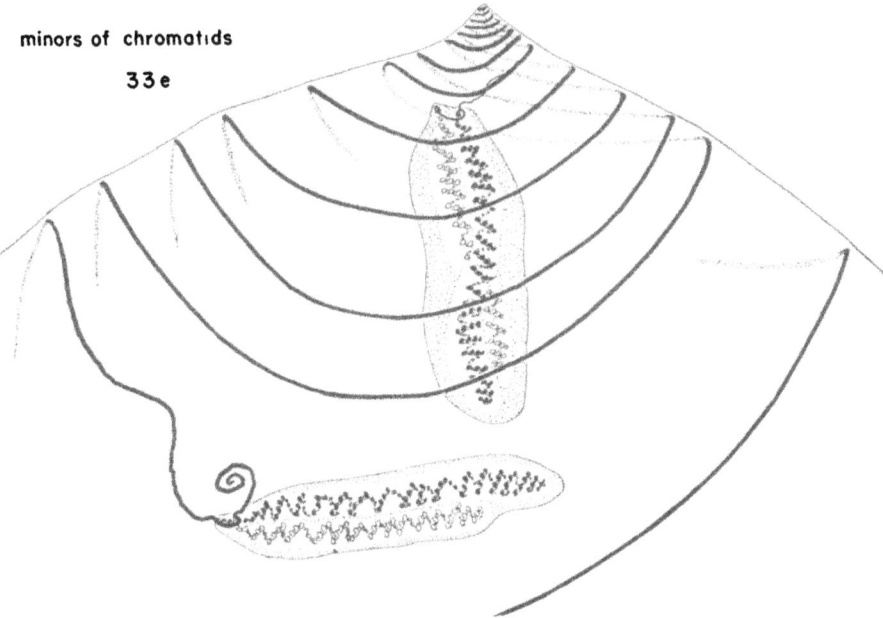

PLATE 18

FIG. 34. Entire cell at the time the fifth flagellar band first becomes separated from the other four bands. × 590.

FIG. 34A. Detail of anterior end of cell showing how band 5 leaves the other bands. The achromatic figure has just pulled apart. These two events must be coordinated; otherwise nuclear division is abnormal. × 1,800.

FIG. 35. The free flagellar band has completed slightly more than half the turns of the body it will have to make in order to separate itself from the other bands. One nucleus and centriole go with this band. The other nucleus and centriole remain with band 4. That portion of the central spindle associated with the anterior or stationary nucleus remains for a considerable time after the portion associated with the moving or posterior nucleus disappears. × 590.

FIG. 36. Flagellar band 5 has completely unwound itself from the other four bands and has partially re-established its coiling system (in the posterior end of the parent cell or what will presently be the anterior end of the new, small, posterior daughter cell).

FIG. 37. The free flagellar band has almost completely reorganized its coiling system, the cytoplasm of the parent cell is constricted preparatory to a transverse division which, when completed, will produce a large anterior daughter cell with four flagellar bands and a small posterior daughter cell with one flagellar band. × 590.

FIG. 38. Anterior daughter cell with four flagellar bands. × 590.

FIG. 39. Posterior daughter cell with one flagellar band. × 590.

PLATE 18

TELOPHASES
band 5 leaving other bands
stages in development of posterior
daughter cell

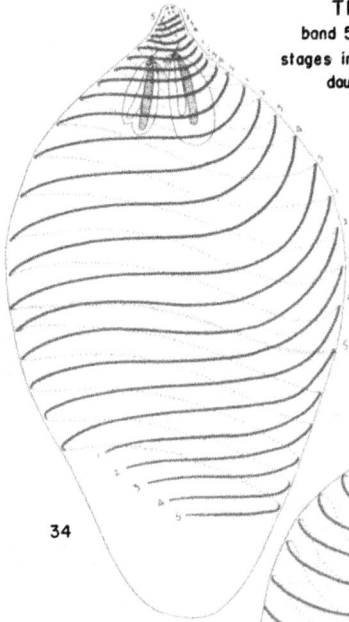

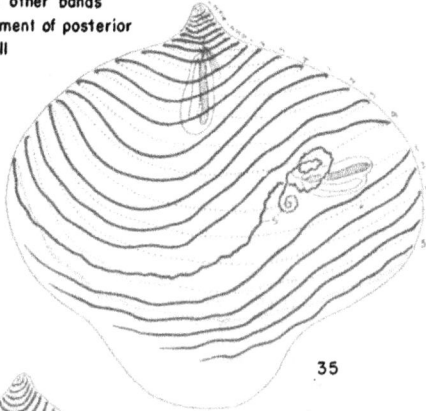

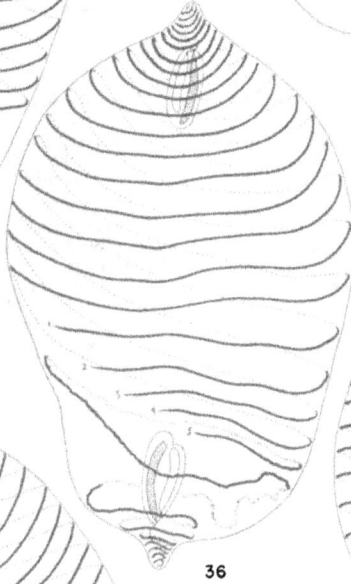

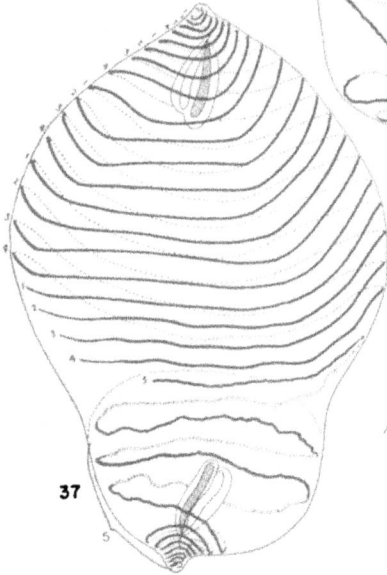

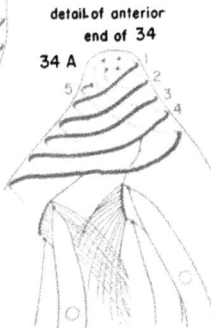

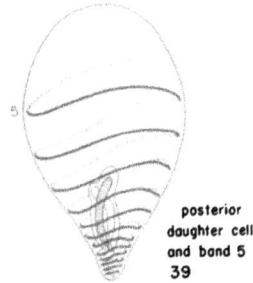

34

35

36

37

anterior
daughter
cell

38

detail of anterior
end of 34

34 A

posterior
daughter cell
and band 5
39

PLATE 19

All figures are of *Holomastigotoides diversa*. Cytoplasmic division is longitudinal, terminal centromeres are anchored to nuclear membrane, and achromatic figure is in contact with nucleus only during metaphase, anaphase, and early telophase.

FIG. 40. Detail of anterior end of cell showing relation of nucleus to achromatic figure in the fairly late prophase which is also the resting stage. Note eight flagellar bands, centrioles which follow bands 4 and 5 for a short distance, become free, and have produced an (resting) achromatic figure from their free ends. × 1,500.

FIG. 40d. A 2-chromosome prophase resting nucleus. Tight, large major coils. Each of the four chromatids has a terminal nucleolus. The two long (dark) chromatids have lateral nucleoli near their centromere ends. The centromeres are permanently anchored to the nuclear membrane but are free to move laterally. × 1,500.

FIG. 41. 3-chromosome telophase nucleus after cytoplasmic division. Two short chromosomes (light) and one long chromosome (dark). Lateral nucleolus of long chromosome fused with terminal of one short chromosome; terminal of long fused with terminal of other short chromosome. Major coils more relaxed than in late prophase (fig. 40d). Centromeres terminal and anchored to nuclear membrane.

FIG. 42. Another type of 3-chromosome telophase nucleus after cytoplasmic division. Two long (dark) chromosomes with terminal and lateral nucleoli and one short (light) chromosome with only a terminal nucleolus. All terminal nucleoli fused. Major coils.

FIG. 43. 2-chromosome fairly early prophase nucleus. A new generation of major coils is forming in each of the four chromatids.

FIG. 44. 2-chromosome prophase nucleus shortly after resting stage. Achromatic figure, which still lies some distance from nucleus, has begun to grow again. Its astral rays will soon make contact with the centromeres of the chromatids (and become chromosomal fibres). Large, tight major coils. Terminal nucleoli of short chromatids fused.

FIG. 45. 4-chromosome diploid nucleus, slightly earlier prophase than previous haploid (fig. 44). Achromatic figure lies farther away from nucleus. Note eight centromeres, one for each chromatid. Between this stage and metaphase, first chromatids will pair, then chromosomes will pair, forming, altogether, two tetrads; one composed of four short homologous chromatids (light); one composed of four long homologous chromatids (dark).

PLATE 19

PROPHASES –
major coils
(40, 43, 44, 45)

40

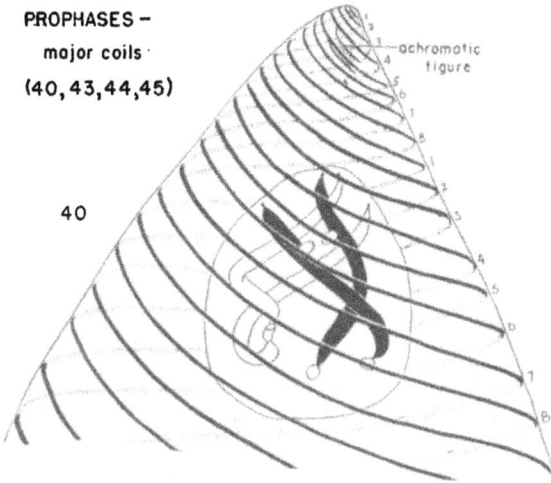

40d

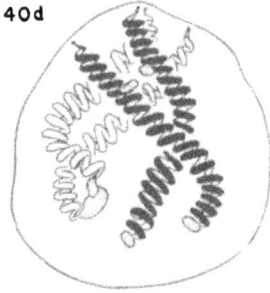

43

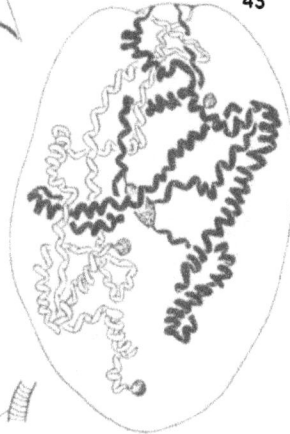

TELOPHASES –
major coils
(41, 42)

41

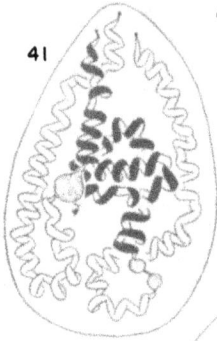

42

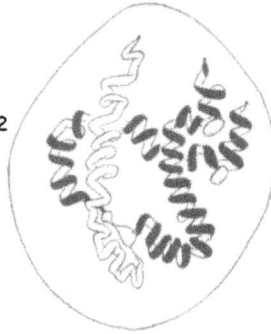

achromatic
figures

haploid   diploid

44   45

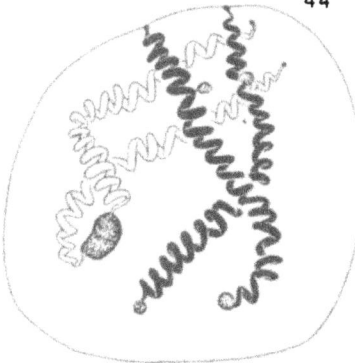

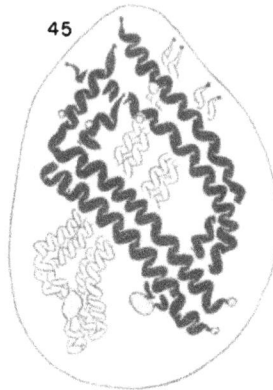

PLATE 20

FIG. 46. Telophase nucleus after cytoplasmic division. Two chromosomes with almost median centromeres, one attached to upper right corner of achromatic figure, other to lower left corner. In the long, darker chromosome, minors are shown incorporated in majors; short chromosome, majors only shown. Matrix of both chromosomes shown. *Holomastigotoides* sp. from *Coptotermes niger* Snyder from Panama.

FIG. 47. Prophase chromatids, median centromeres, one attached to each corner of the achromatic figure. Note the very large, loose major coils; in the two longer, darker chromatids minors are shown incorporated in majors; in short ones only majors shown. This is a 2-chromosome *Holomastigotoides* sp. from *Psammotermes fuscofemoralis* from Egypt. Delafield's hematoxylin, Schaudinn's fixative.

FIG. 48. Prophase chromatids, two on left have terminal centromeres attached to upper and lower corners of achromatic figure; two on right have median centromeres attached to upper and lower corners of achromatic figure. The chromatid on the left, minors in majors and the beginning of supers are shown. The one next to it on right, supers are better developed. The long, darker chromatid on right has three well-developed supers in right hand arm and the beginning of supers in the other arm. Its sister, 2-armed chromatid (which has been drawn light), has two well-developed super coils in each arm. 2-chromosome *Holo-mastigotoides* sp. from *Coptotermes niger* Snyder from Panama.

FIG. 49. Prophase chromatids of 2-chromosome *Spirotricho-nympha mirabilis* from *Porotermes adamsoni* Froggatt from A.C.T., Canberra, Australia. Major coils only shown in two chromatids on left and both majors and minors shown in those on right. Centromeres median and submedian.

FIG. 50. Telophase chromatids shortly before cytoplasmic division. Centromeres nearly median. Chromatid on left only super coils shown in each arm. The left arm of the other chromatid has minors, majors, and supers shown; the right arm has only majors and supers shown. 2-chromosome *Spirotrichonympha mirabilis* from *Porotermes adamsoni* Froggatt from A.C.T., Canberra, Australia.

FIG. 51. This is a metaphase of the same organism shown in figure 48. Two chromatids on left have terminal centromeres attached to upper and lower corners of the achromatic figure: in the right one, supers with majors incorporated in them shown; in left one, minors incorporated in majors, and majors incorporated in supers shown. The other two chromatids have median centromeres attached to the upper and lower right corners of the achromatic figure: in left one, only supers shown in both arms; in right one, left arm has majors incorporated in supers; in right arm, minors, majors, and the beginning of supers shown.

66

PLATE 20

TELOPHASE
chromosomes

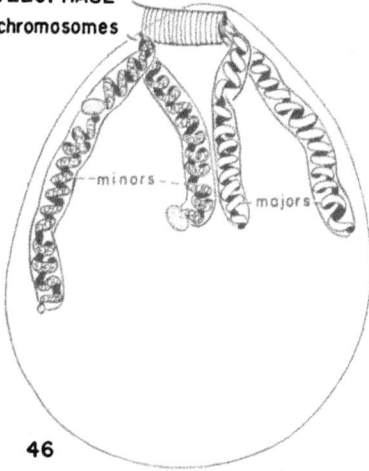

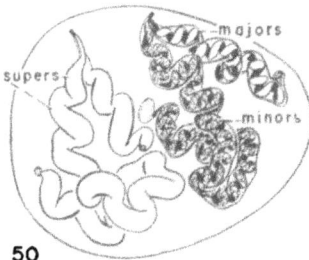

PROPHASE
chromatids

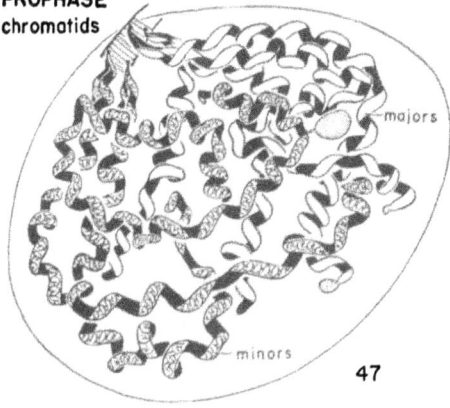

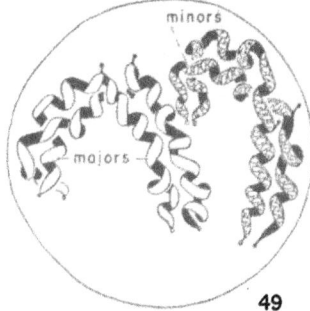

minors

majors

majors

minors

47

46

late PROPHASE chromatids

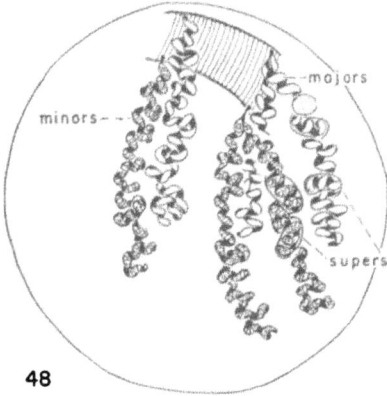

majors

minors

supers

48

PROPHASE chromatids

minors

majors

49

METAPHASE chromatids

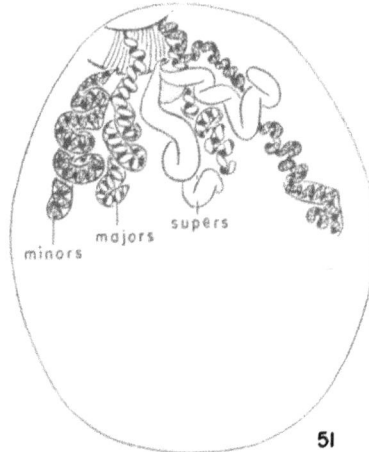

minors

majors

supers

51

TELOPHASE chromosomes

majors

supers

minors

50

PLATE 21

FIG. 52. *Trichomonas* [= *Pseudotrypanosoma*] *gigantea* from *Porotermes adamsoni* Froggatt from A.C.T., Canberra, Australia. Anaphase chromatids with median and near median centromeres. Chromatids on right, major coils only shown; those on left, minors incorporated in majors. Note very large astral rays characteristic of this and closely related species. Distal ends of the elongate centrioles are bent so that the central spindle portion of the achromatic figure which they produce is nearly a complete cylinder. No centrosomes on distal ends of centrioles.

FIG. 53. *Trichomonas* sp. from *Porotermes adamsoni* Froggatt from A.C.T., Canberra, Australia. Telophase after nuclear division. Only major coils shown in chromatids in right hand nucleus; both minors and majors shown in those of left hand nucleus. In this genus, the nuclear membrane remains intact during division.

PLATE 21

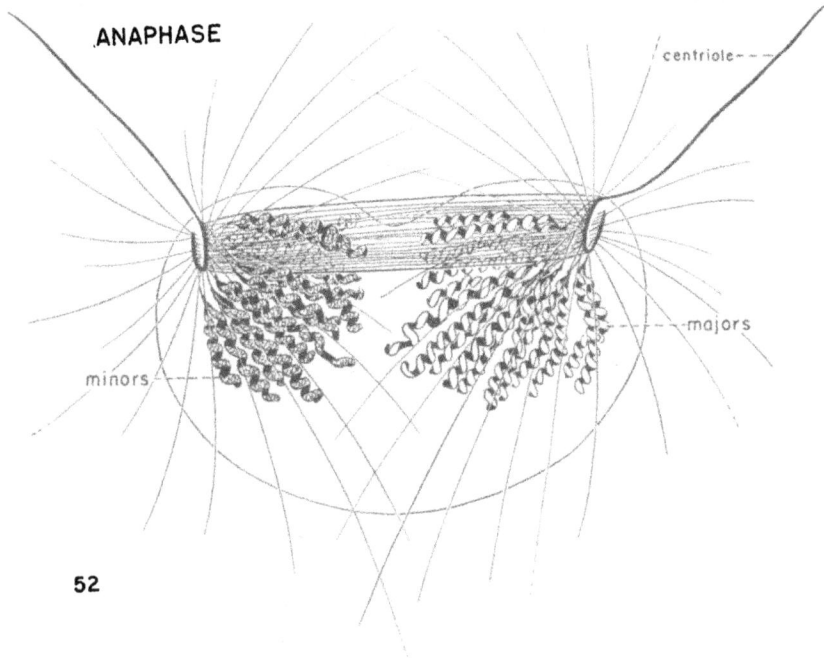

ANAPHASE

52

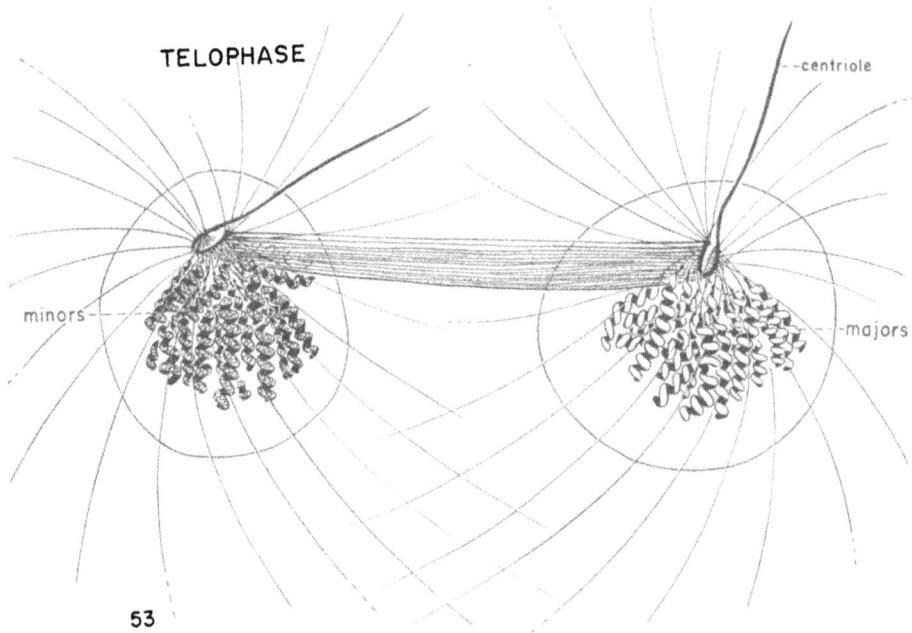

TELOPHASE

53

PLATE 22

Fig. 54. Telophase after nuclear division of *Joenina pulchella* from *Porotermes adamsoni* Froggatt from A.C.T., Canberra, Australia. Centromeres are subterminal and median. Only majors shown in chromatids of upper nucleus; majors and minors shown in most of those in lower nucleus. In this and related genera of the Lophomonadidae centrioles are not elongate; fibres of the achromatic figure arise from practically all of their surface; they have no centrosomes. The new centriole produces extranuclear organelles in this generation and, together with the old one beside it, will take part in producing an achromatic figure in the next generation.

Fig. 55. A single telophase chromatid of *Barbulanympha* from *Cryptocercus punctulatus.* Centromere is median or nearly so.

Fig. 56. A telophase chromatid, with median centromere, from *Pseudotrichonympha* sp. from *Prorhinotermes simplex* Hagen from Coral Gables, Florida.

Fig. 57. A telophase chromatid from *Oxymonas* [= *Saccinobaculus*] *doroaxostylus* from *Cryptocercus punctulatus.*

PLATE 22

TELOPHASES

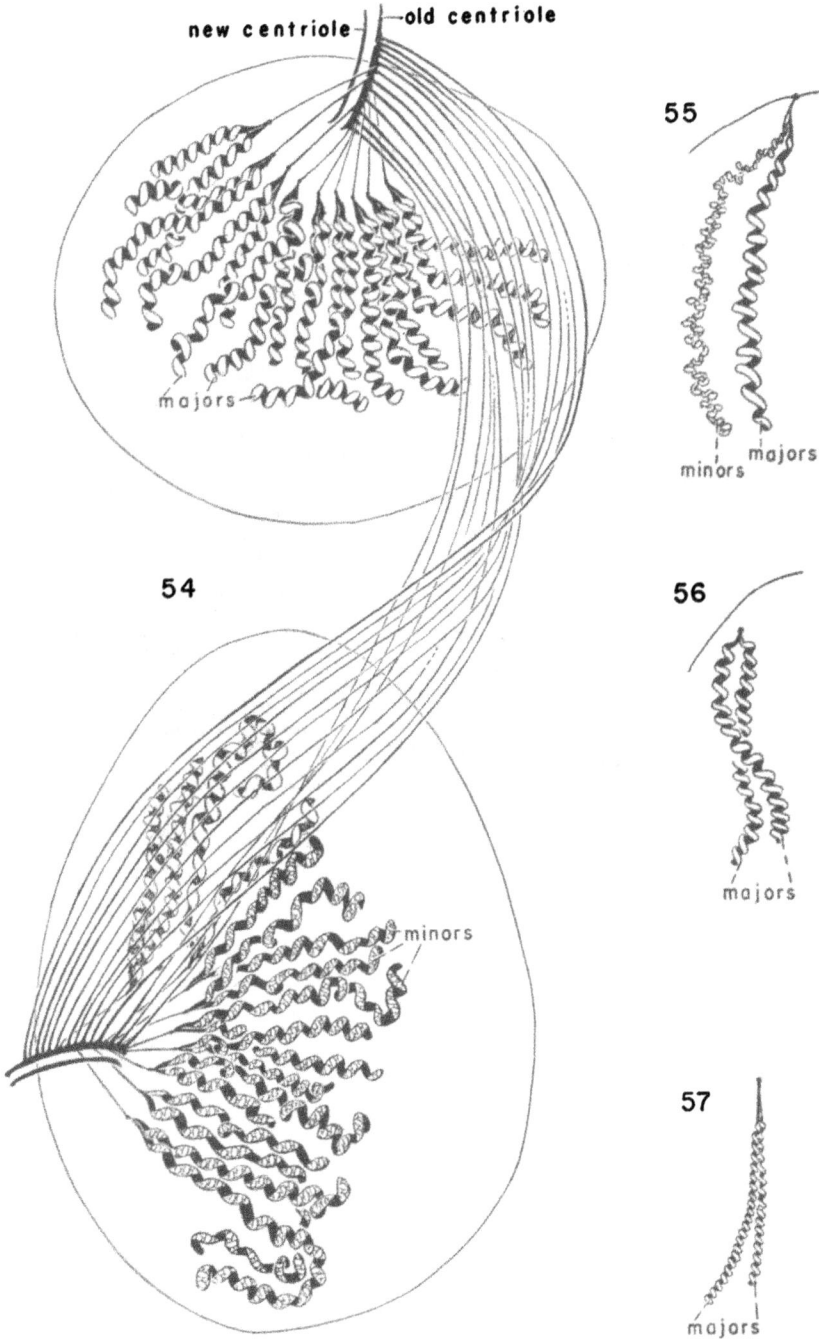

PLATES 23–34

DOUBLE CHROMATID CHROMOSOMES

PLATE 23

FIG. 58. · Telophase chromosomes after cytoplasmic division and before twisting has begun. · Both chromosomes are anchored to the free, distal end of the old centriole which, for most of its length, follows flagellar band 4. The two chromosomal membranes are fusing to form a long, narrow, new nuclear membrane. Pieces of old terminal nucleolus within membrane surrounding short chromosome.

FIG. 58b. Five gyres of relational coiling of the short (light) chromatids and five of the long (dark) chromatids. Terminal nucleoli of sister chromatids fused; those of short chromatids just beginning to form.

FIG. 58d. Individual major coils of each chromatid are very close together. In double-chromatid chromosomes, there are never any major coils of chromosomes. ·

FIG. 58e. A small portion of the anterior or centromere end of each chromatid showing that small, minor coils are incorporated in each major coil. × 7,200.

FIG. 59a. Individual twisting of chromosomes has begun. Beginning stage of collapse of matrix. Chromosomes now in common nuclear membrane. New flagellar band 5 and new centriole are growing out.

FIG. 59b. Relational coiling of chromatids of both short and long chromosome—four gyres in short one and six in long one. All terminal nucleoli fused; laterals of long chromatids fused.

FIG. 59d. Chromatids appear longer but their major coils are still very close together.

FIG. 60a. More individual twisting of chromosomes. Nucleus has become wider and shorter to accommodate changes in chromosomes.

FIG. 60b. Relational coiling of chromatids looser but none has come out; six gyres in short sister chromatids and nine in long ones. The fact that more gyres are present here than in the two earlier stages is merely a matter of individual variation. Terminal nucleoli of sister chromatids fused; laterals of long ones fused. New achromatic figure has formed.

FIG. 60d. Major coils considerably looser due to collapsing of matrix in which each chromatid is embedded. Duplication of centromeres has occurred and the centromere ends of the chromatids have separated, only one chromatid in each sister pair moving. Duplication of chromatids occurred in previous early prophase but centromere duplication, unlike that of single-chromatid chromosomes, does not occur until now.

72

PLATE 23

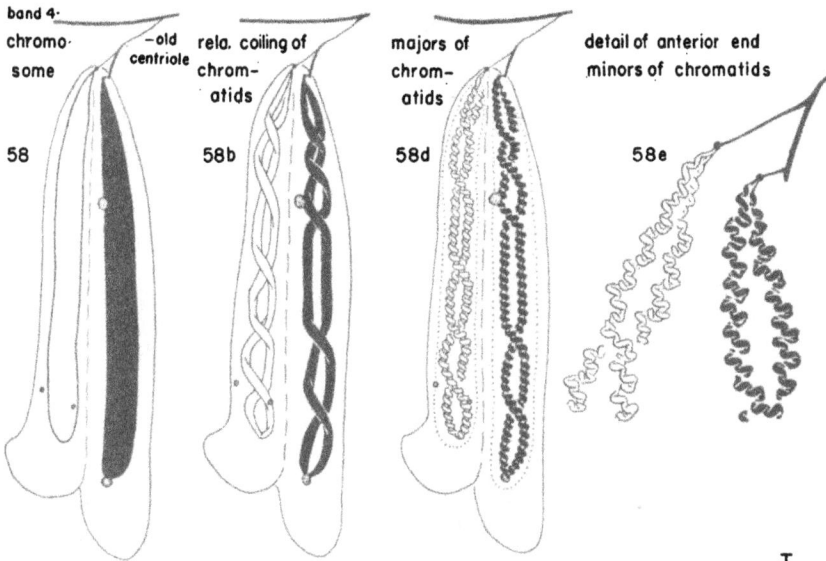

band 4·
chromo·
some

–old
centriole

58

rela. coiling of
chrom–
atids

58b

majors of
chrom–
atids

58d

detail of anterior end
minors of chromatids

58e

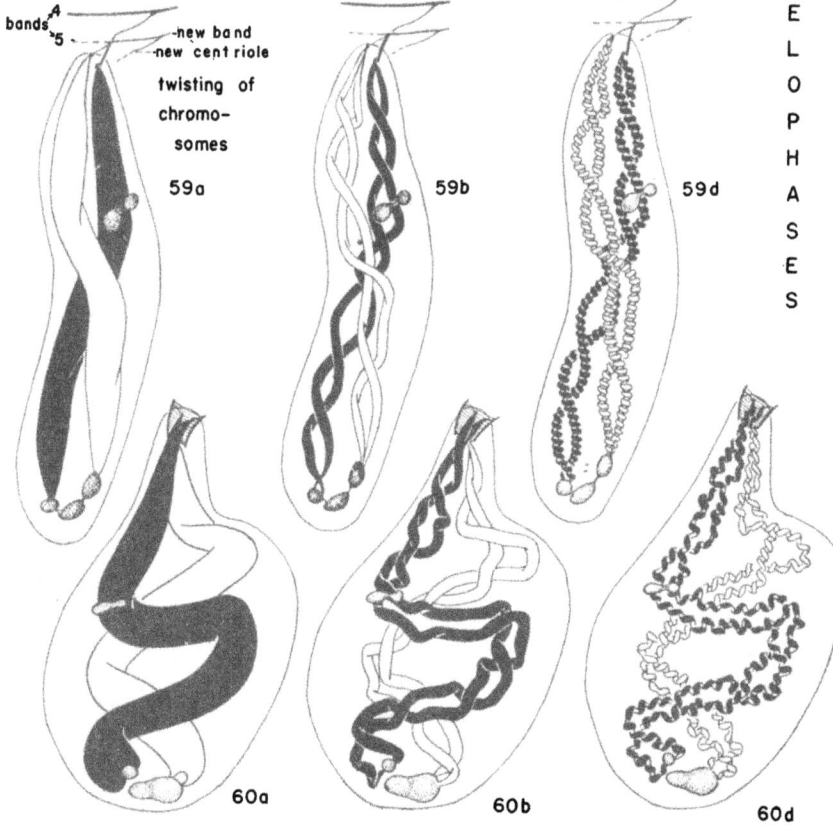

A
bands
5

–new band
–new centriole

twisting of
chromo–
somes

59a

59b

59d

60a

60b

60d

T
E
L
O
P
H
A
S
E
S

PLATE 24

FIG. 61a. More progress in twisting of chromosomes.

FIG. 61b, d. Relational coiling of chromatids still present. Major coils very loose all along each chromatid. They have all but disappeared and will not be shown again until a new generation is produced in the prophase (by contraction of the matrix).

FIG. 61e. Detail of centromere ends of each chromatid showing minor coils in the greatly relaxed, disappearing majors. × 7,200.

FIG. 62a. Much individual twisting of chromosomes.

FIG. 62b. Relational coiling of chromatids much looser but none has come out.

FIG. 62c. Duplication of chromatids has occurred. Each chromatid is now composed of two half chromatids. The half chromatids have very many gyres of tight relational coiling. Two generations of relational coiling are now present, that of chromatids which was produced at this stage (early prophase) in the last generation, and that of the half chromatids which has just been produced. Note that centromeres in double-chromatid chromosomes are not duplicated when the chromatids are. This does not occur until late telophase.

FIG. 62e. Detail of centromere ends of half chromatids (eight in all) showing their minors. No majors are present at this stage. The bending of the chromonemata of each half chromatid is produced by the tight relational coiling of the half chromatids. × 7,200.

PLATE 24

TELOPHASES

twisting of chromo- somes

6la

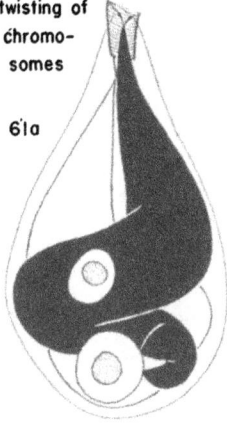

rela. coiling of chromatids and remnants of major coils

6lb,d

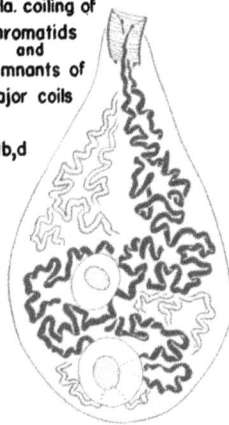

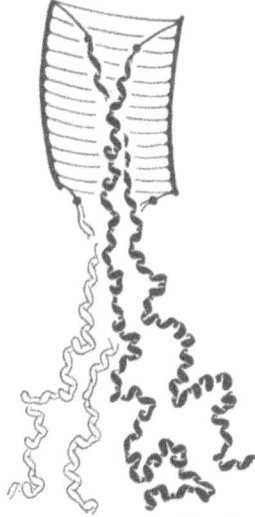

6le   details of anterior end minors of chromatids

PROPHASES

twisting of chromosomes

62a.

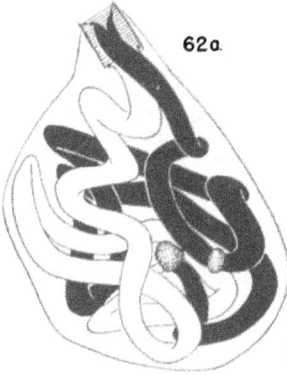

rela. coiling of chromatids

62b

rela. coiling of ½ chromatids

(chromatids duplicated)

62c

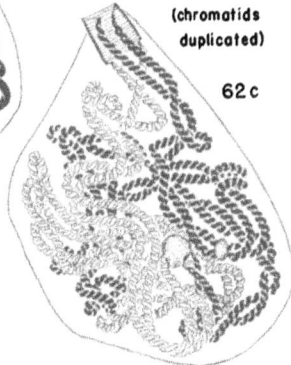

detail of anterior end minors of ½ chromatids

62e

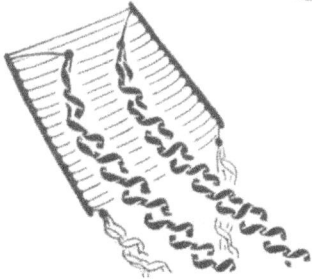

PLATE 25

FIG. 63a. Chromosomes still greatly twisted individually.

FIG. 63b. Relational coiling of chromatids still very evident in places—eight gyres in short sister chromatids, seven in long ones. Terminal nucleoli of each pair of sister chromatids fused; lateral nucleoli fused.

FIG. 63c. Very many gyres of relational coiling of sister half chromatids. Considerable progress in unwinding has been made in some places while little or no progress has been made in others. Unwinding can begin at any point along the half chromatids.

FIG. 63e. Detail of centromere ends of half chromatids showing their minor coils which have begun to tighten in places preparatory to the production of a new generation of major coils. × 7,200.

FIG. 64a. Individual twisting of chromosomes and two gyres of relational twisting.

FIG. 64b. Little progress has been made in unwinding of relational coils of chromatids.

FIG. 64c. However, very great progress has been made in unwinding of the relational coiling of the sister half chromatids. They, as well as the chromatids and chromosomes, now appear much shorter than when duplication occurred.

FIG. 64d. Contraction of the matrix has caused the chromonema of half chromatids to bend on itself and thus begin a new generation of major coils in which the ever present minors are incorporated. These early, loose majors, like the later tight ones, are smaller than those of single-chromatid chromosomes (see figs. 17d–20d).

PLATE 25

PROPHASES

twisting of
chromosomes
63a

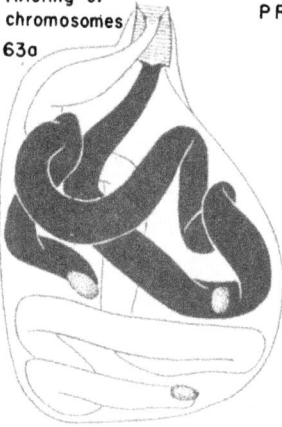

rela. coiling of
chromatids
63b

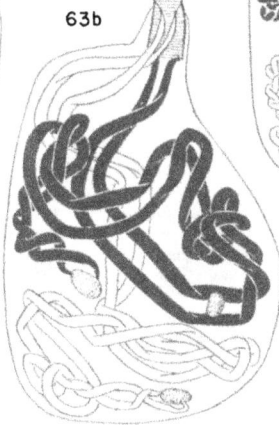

rela. coiling of
½ chromatids
63c

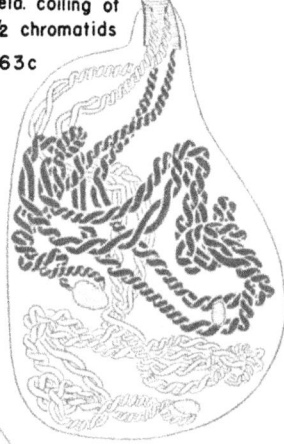

twisting of
chromo-
somes
64a

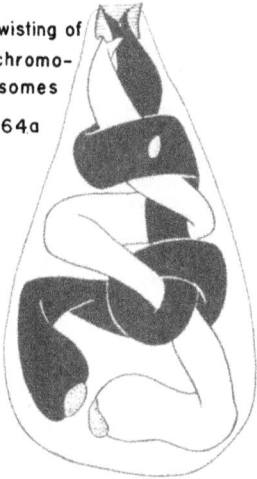

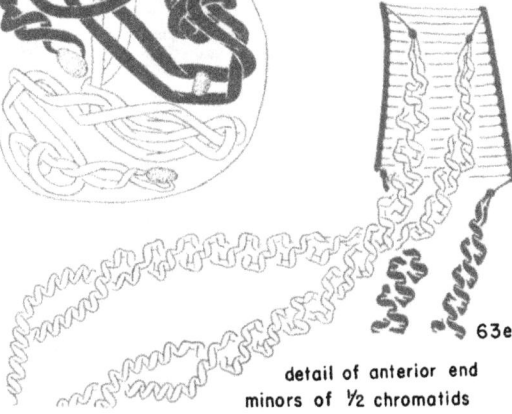

63e

detail of anterior end
minors of ½ chromatids

rela. coiling of
chromatids
64b

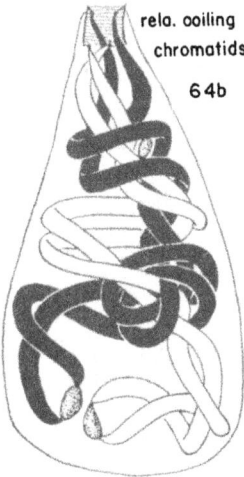

rela. coiling of
½ chromatids
64c

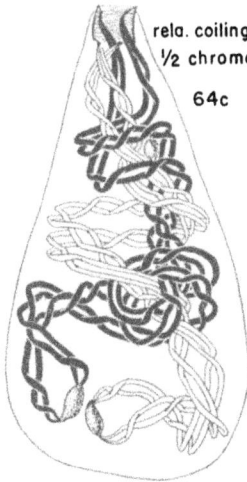

majors of ½
chromatids
64d

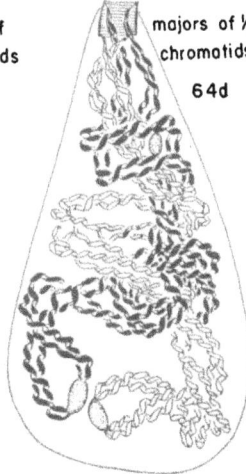

PLATE 26

Fig. 65a. Still a considerable amount of relational and individual twisting of chromosomes.

Fig. 65b. Relational coiling of sister chromatids is coming out —two gyres in short ones, three in long ones. Terminal and lateral nucleoli of sister chromatids fused.

Fig. 65c. Progress has also been made in unwinding of relational coiling of sister half chromatids. There are now long stretches with only a few gyres.

Fig. 65d. The major coils of each half chromatid have become considerably tighter and have thus made the half chromatids appear shorter.

Fig. 65e. Detail of centromere ends of half chromatids showing their minor coils. × 7,200.

Fig. 66a. Chromosomes appear much shorter due to loss of much of their individual twisting. About a gyre of relational twisting remains.

Fig. 66b. Relational coiling of sister chromatids is out except for slightly more than one gyre.

Fig. 66c. Relational coiling of sister half chromatids is coming out rapidly too but all of it will not come out before contraction of matrix stops and thus prevents further loss of this coiling until the next generation when contraction of the matrix again produces rotation and unwinding of these half chromatids (they will be chromatids then). Nucleoli of half chromatids fused; those of chromatids not fused.

Fig. 66d. Major coils of half chromatids are becoming quite tight and close together.

PLATE 26

PROPHASES

twisting of chromo- somes

65a

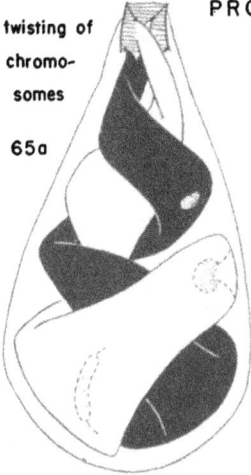

rela. coiling of chromatids

65b

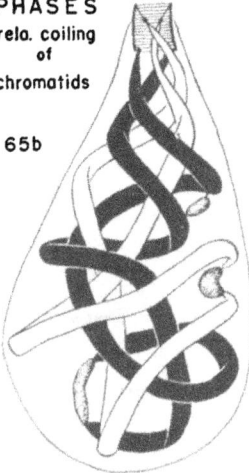

rela. coiling of ½ chrom- atids

65c

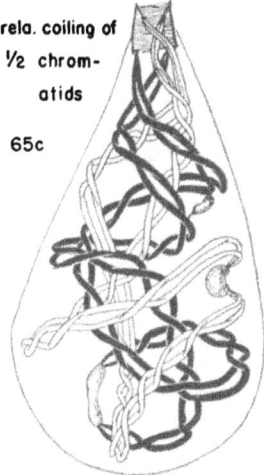

majors of ½ chrom- atids

65d

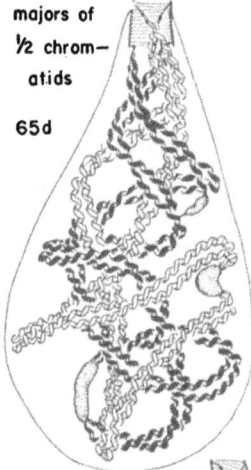

detail of anterior end minors of ½ chromatids

65e

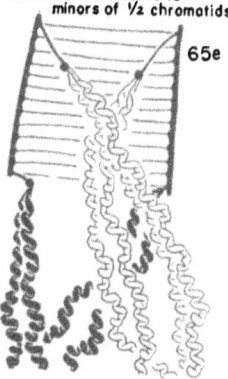

twisting of chromo- somes

66a

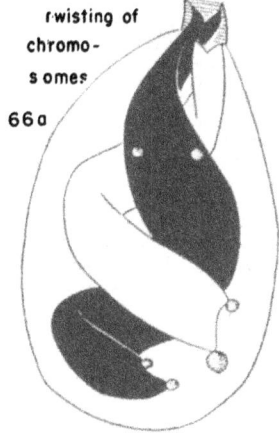

rela. coiling of chrom- atids

66b

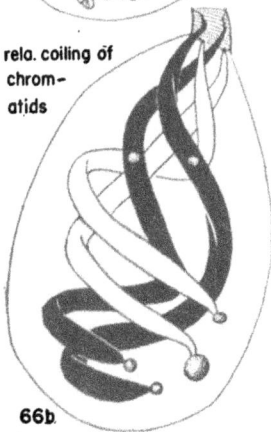

rela. coiling of ½ chrom- atids

66c

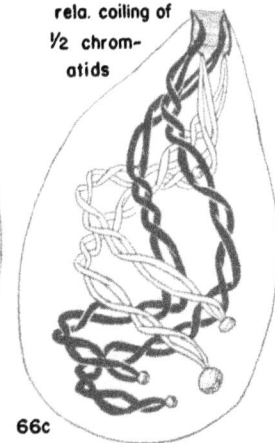

majors of ½ chrom- atids

66d

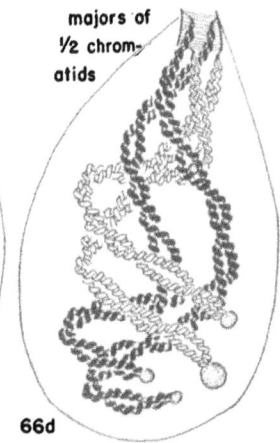

PLATE 27

Fig. 67a. Less than one gyre of chromosomal twisting left.

Fig. 67b. Slightly over a gyre of relational coiling of sister chromatids left.

Fig. 67c. Relational coiling of sister half chromatids unwound about as much as will occur in this generation: six gyres in one short pair of sisters, five in other; five gyres in each pair of long sisters. Sometimes a few more, sometimes a few less gyres of relational coiling than this goes through metaphase, anaphase, and telophase to the prophase of the next generation before it comes out. Nucleoli of half chromatids fused; those of chromatids not fused. This is the resting stage, the same place where development in single- as well as double-chromatid stops, for several days, sometimes several weeks.

Fig. 67d. The majors of each half chromatid are now quite tight. In the prophase development following the resting stage, they will tighten a little more due to further contraction of the matrix which then removes all of the relational coiling of chromatids.

Fig. 68a. Early prophase following resting stage. Very small amount of chromosomal twisting remains. Presently it will disappear entirely.

Fig. 68b. Less than a gyre of relational coiling of chromatids remains. Thus ends a generation of relational coiling which began (as half chromatid relational coiling) in the very early prophase of the previous generation. It persisted from very early prophase of one generation through all stages of that generation and to the end of the prophase of the next generation.

Fig. 68c. Only three gyres of sister half chromatid relational coiling left. Sometimes as many as ten gyres are left at this stage. No more come out until the prophase of the next generation.

Fig. 68d. Major coils of the half chromatids are as tight as they will ever be. If more contraction of the matrix were to occur, the chromonema would bend and thus begin the development of super-coils, as occurs in some species of *Holomastigotoides*. Nucleoli of sister half chromatids fused.

Fig. 68e. Detail of centromere ends of half chromatids showing their minor coils incorporated in majors. $\times 7{,}200$.

PLATE 27

PROPHASES

twisting of
chromo-
somes

67a

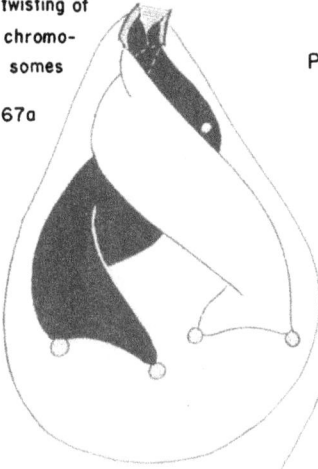

67c
rela. coiling of
½ chromatids

rela. coiling of
chromatids

67b

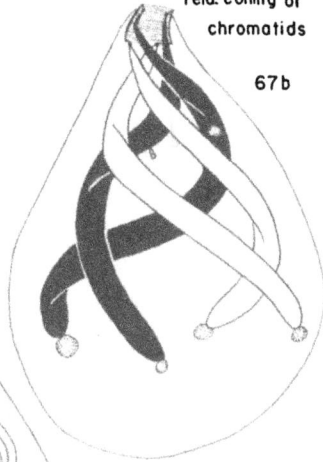

majors of
½ chrom-
atids

67d

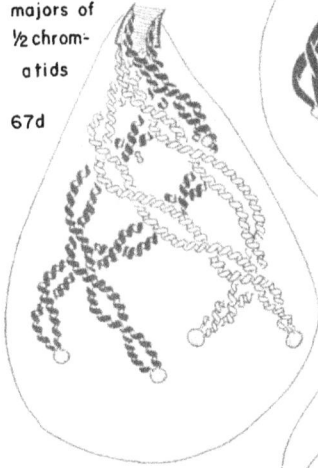

twisting
of
chromo-
somes

68a

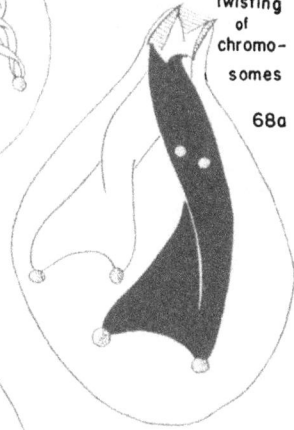

rela. coiling of
½ chrom-
atids

68c

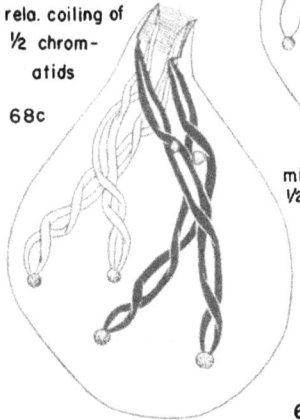

68b

rela. coiling of
chromatids

minors
½ chit'ds

68e

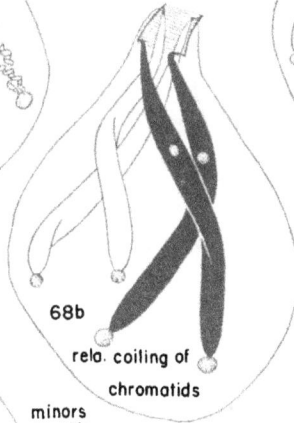

majors of
½ chrom-
atids

68d

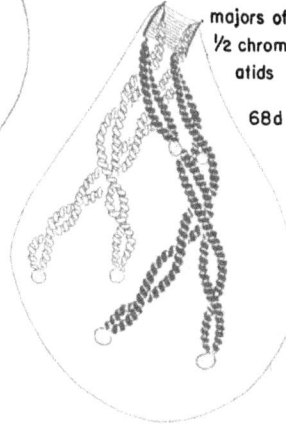

PLATE 28

FIG. 69d. Majors of half chromatids very close together. Less relational coiling of sister half chromatids than in next illustration which is a slightly later stage. Very earliest stage in segregation of the small amount of nuclear material which will remain with the chromatids from the large amount that will be discarded into the cytoplasm.

FIG. 70d. Tight major coils of sister half chromatids. More than usual number of gyres of relational coiling of sister half chromatids. All nucleoli have disappeared; very small new ones forming on distal ends of half chromatids. They fuse almost as soon as formed. More progress in demarkation of nuclear material to be retained from that to be discarded into cytoplasm. Some nuclear material already being thrown into cytoplasm. Achromatic figure has begun to grow again. As it increases in length, the sister chromatids are separated, the separation always progressing posteriorly from the centromere ends.

FIG. 71d. No change in major coils and relational coiling of .

half chromatids. Achromatic figure larger, sister chromatids separated, two non-sisters (light and dark) lie together. More nuclear material has been discarded. New nucleoli have been produced and the terminal sister ones of half chromatids have fused. Remains of nucleolar material still in nucleus will be discarded presently into cytoplasm where it may be followed (remains distinct) for some time in living cells. This may be called either anaphase or metaphase.

FIG. 72d. Central spindle has become longer, is bent, and the chromatids have become segregated into more distinct, developing, daughter nuclei. Flagellar band 5 has separated from the other four bands anteriorly and has begun the slow, complicated process of turning round and round the cell and thus unwinding itself completely from the other bands. This is an example, which one sometimes sees, of the fifth flagellar band beginning to free itself from the other somewhat prematurely. However, unless it frees itself and unwinds much too early, an abortive, premature attempt at nuclear division does not result (has been seen only once).

PLATE 28

Majors of ½ Chromatids
(69-72)

PROPHASE

69d

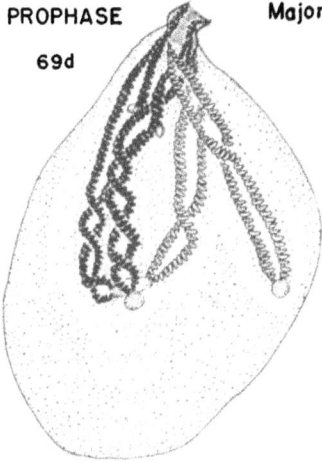

PROPHASE

70d

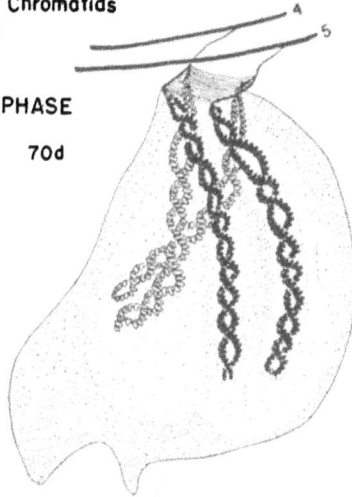

METAPHASE   71d

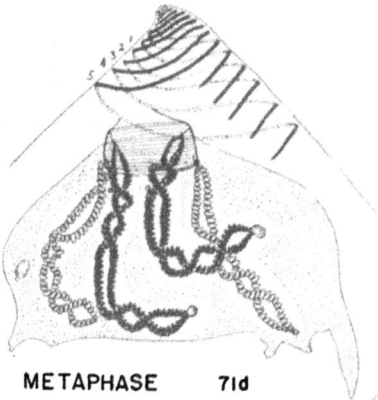

TELOPHASE

72d

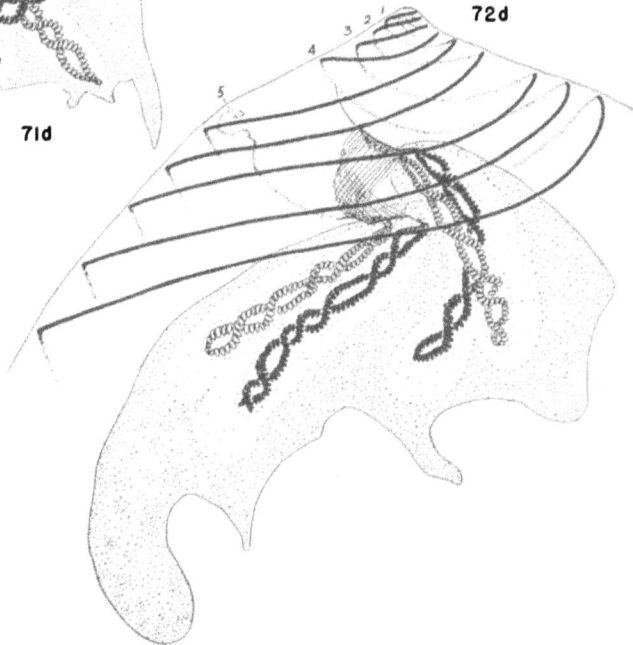

PLATE 29

FIG. 73. Here are the four chromatids, two non-sisters in each group, drawn as they appear when greatly overstained. One cannot see that each is composed of two relationally coiled half chromatids and that each half chromatid has individual major and minor coils. Central spindle is greatly bent.

FIG. 73c. Here one sees the two relationally coiled sister half chromatids of each chromatid.

FIG. 73d. Here one sees the tight major coils of each half chromatid. Nucleus has discarded a good deal of its material into the cytoplasm. The half chromatids are beginning to produce new terminal nucleoli as indicated by the joining, in some instances, of their distal ends.

FIG. 73e. Detail of centromere ends of two pairs of half chromatids (sisters and non-sisters). Here one sees that each major coil of each half chromatid has minor coils incorporated in it. × 7,200.

FIG. 74d. Central spindle has pulled apart. Pieces of nuclear material are breaking away and becoming free in the cytoplasm. Clear demarkation of nuclear material which will and which will not be discarded into cytoplasm. Note here, and also in earlier figures, that the membrane separating the nuclear material to be discarded from that to be retained forms around each chromatid rather than around each half chromatid. In this respect, and also in the number of centromeres present, single- and double-chromatid chromosomes are alike.

PLATE 29

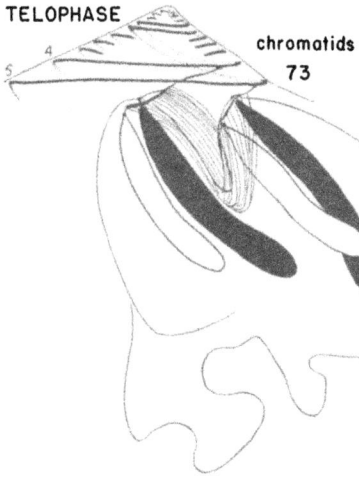

TELOPHASE

chromatids

73

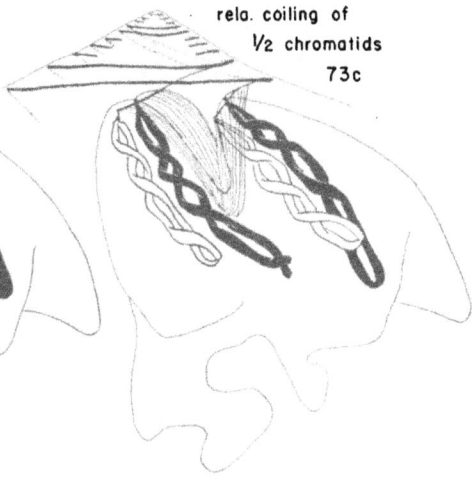

rela. coiling of
½ chromatids

73c

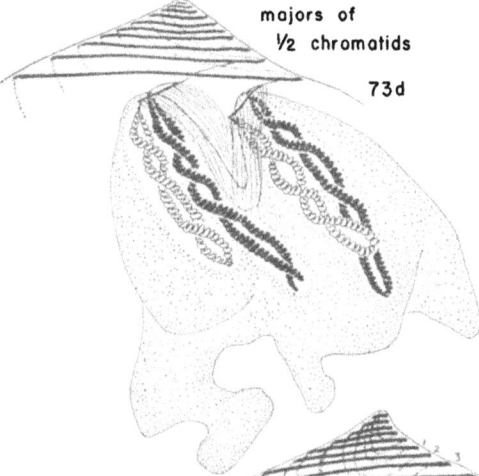

majors of
½ chromatids

73d

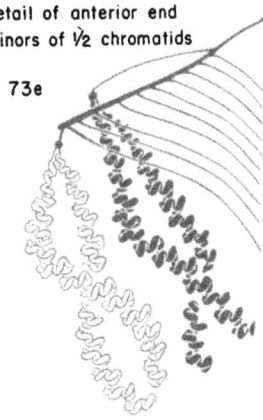

detail of anterior end
minors of ½ chromatids

73e

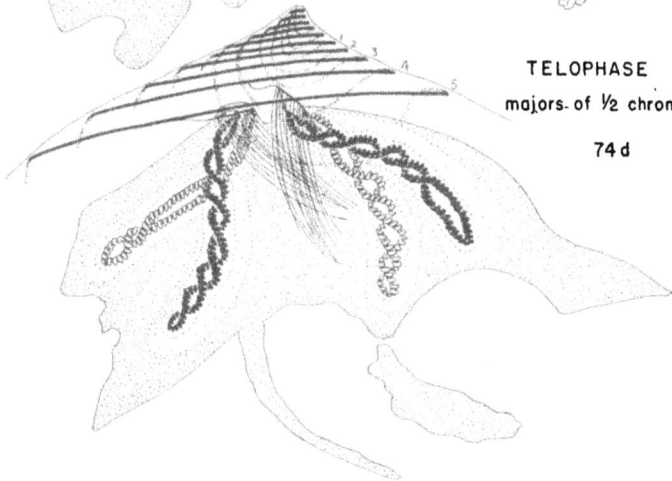

TELOPHASE

majors of ½ chromatids

74d

PLATE 30

FIG. *75d.* Remains of each half of the achromatic figure, one half going with each daughter nucleus. Note the great difference in size of the nuclei. The small one, by means of the remains of the achromatic figure, is anchored to the centriole that follows the fifth flagellar band. This will be the nucleus of the small, posterior daughter cell. It always discards its nuclear material into the cytoplasm earlier than the other nucleus. Pulling away of the fifth flagellar band is slightly behind schedule. Major coils as in previous figures.

FIG. *76d.* Fifth flagellar band behaving as it usually does at a given stage in nuclear development. In other words, behavior of both nucleus and flagellar band are properly coordinated. Band has made about two turns in unwinding itself from other bands. Remains of central spindle produced by centriole that follows band 4 present; nothing left of that produced by centriole that follows band 5. This is usual behavior. No change in major coils.

FIG. *77d.* More progress in unwinding by flagellar band 5; other features much as in two previous figures.

PLATE 30

75d

76d

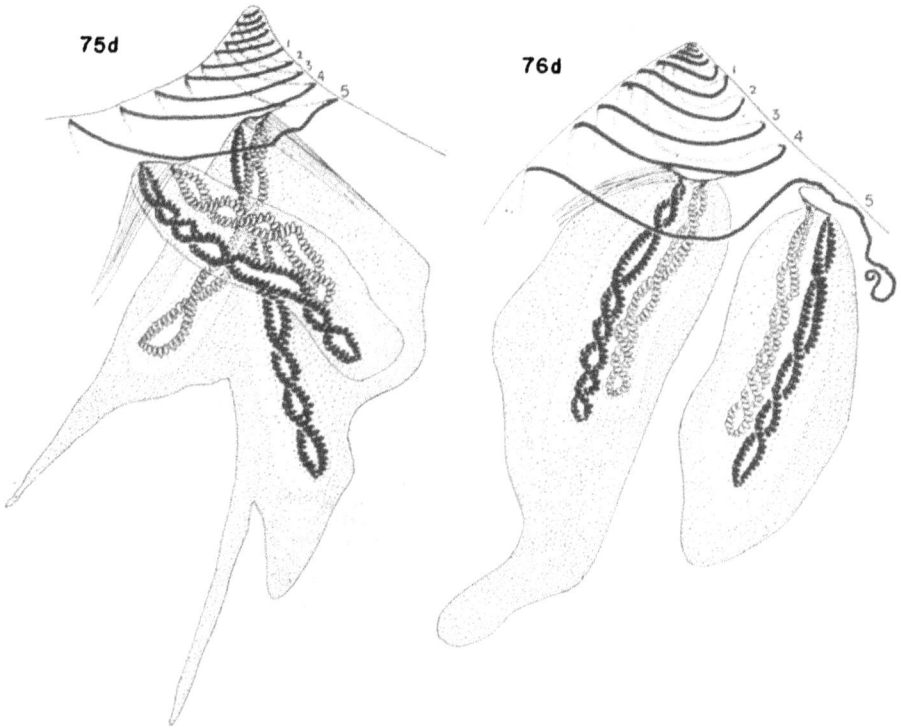

TELOPHASES
Majors of ½ Chromatids (75-77)

77d

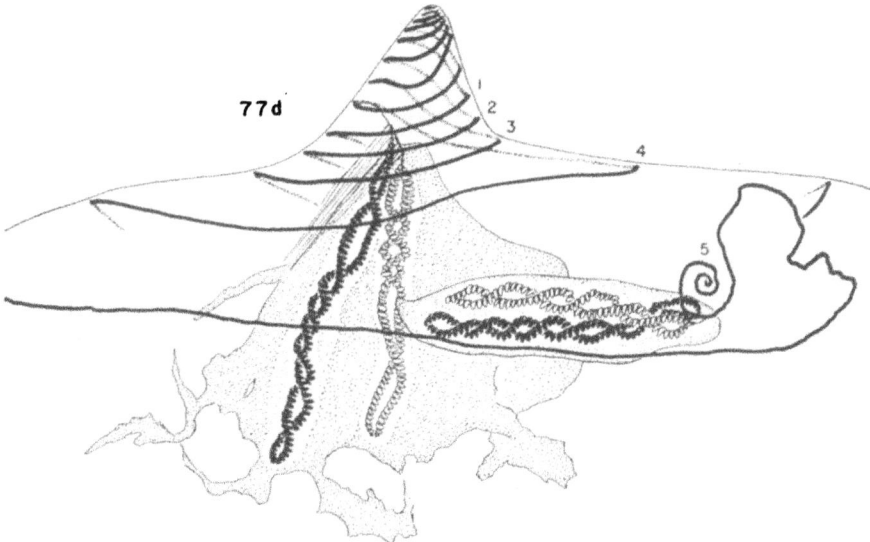

PLATE 31

Fig. 78. Entire cell at low magnification to show progress in unwinding of flagellar band 5. × 590.

Fig. 78c. Relational coiling of sister half chromatids, in posterior and anterior daughter nuclei, has not changed since the late prophase when contraction of the matrix of each half chromatid stopped.

Fig. 78d. The major coiling of the sister half chromatids also has not changed at all since the late prophase.

PLATE 31

TELOPHASE

rela. coiling of ½ chromatids

78c

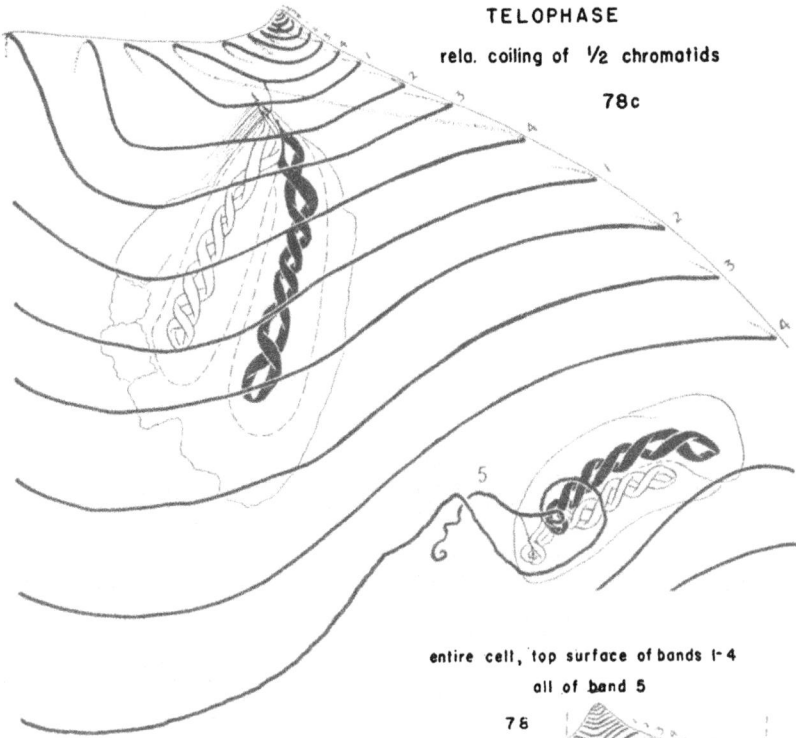

entire cell, top surface of bands 1-4

all of band 5

78

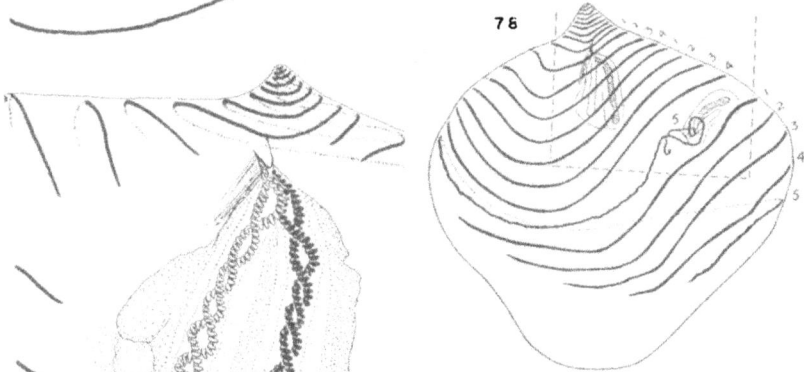

←majors of ½ chromatids→

78d

anterior nucleus

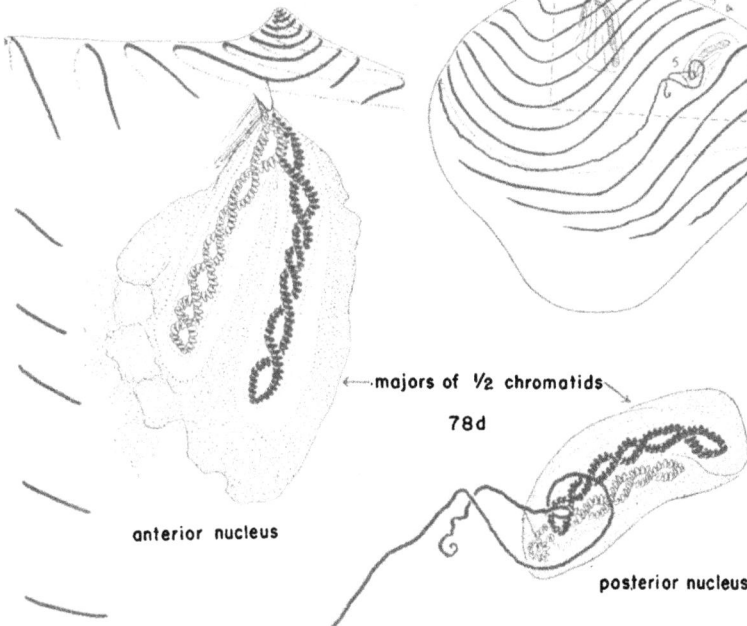

posterior nucleus

89

PLATE 32

FIG. 79. Entire cell at low magnification. The fifth flagellar band has almost completed its seemingly difficult task of unwinding and separating itself from the other four bands. × 590.

FIG. 79d. Major coils of nuclei of developing posterior and anterior daughter cells still as in late prophase. The nucleus of the developing posterior daughter cell is suspended by the centriole which, for most of its length, follows flagellar band 5. The membranes surrounding the chromatids of this nucleus are fusing to form a new nuclear membrane.

FIG. 80. Entire cell at low magnification. Further progress in unwinding of flagellar band 5 and recoiling of this band in the posterior end of the parent cell (this will be the anterior end of the posterior daughter cell after cytoplasmic divi-

sion). Constriction of the cytoplasm preparatory to unequal, transverse division of the cell, producing a large, anterior-daughter and a small, posterior daughter cell (see figs. 38, 39). × 590.

FIG. 80d. No change in major coils since late prophase. Presently cell division will occur and the major coils will continue unchanged to the telophase after cytoplasmic division stage (fig. 58d) where the illustrations of them began. In the posterior nucleus, old, discarded nucleoli and developing new ones are present. In the anterior nucleus, new terminal nucleoli must be developing; otherwise the ends of the half chromatids would not be fusing; a portion of one nucleolus is about to be discarded from the nucleus (on the right) and a discarded nucleolus lies in the cytoplasm to the left of the nucleus.

90

PLATE 32

TELOPHASES
entire cells

80

79

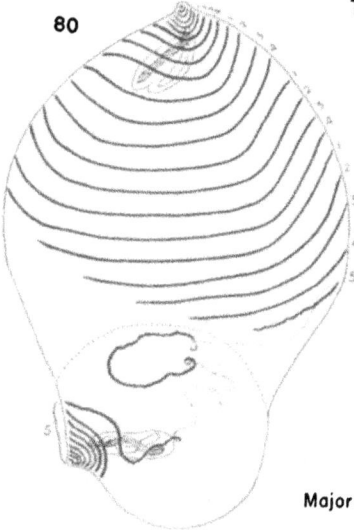
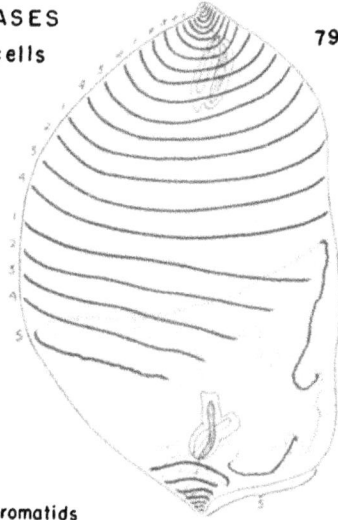

Majors of ½ Chromatids

anterior nuclei

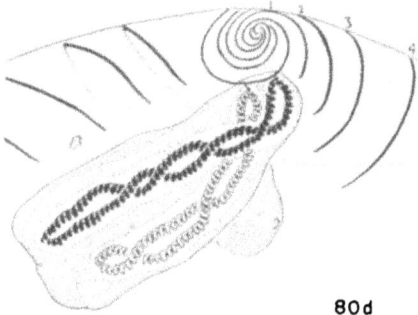
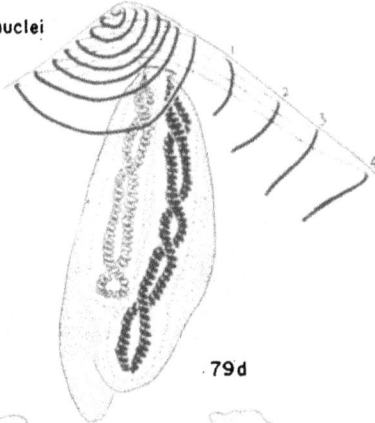

80d

79d

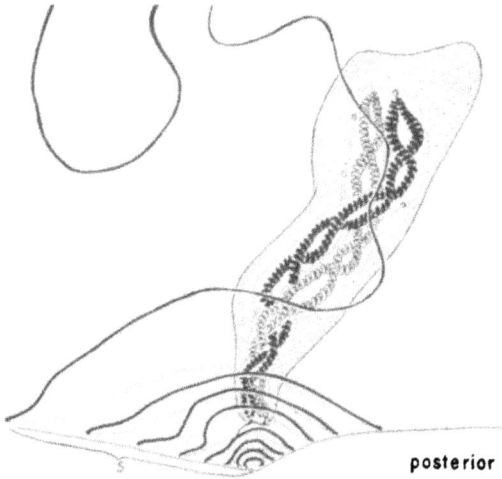
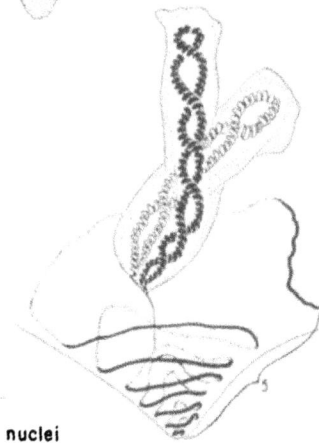

posterior nuclei

PLATE 33

FIG. 81. Telophase after cytoplasmic division of a 3-chromosome variety of *H. tusitala*. Two long (dark) chromosomes with lateral and terminal nucleoli and one short (light) chromosome with only a terminal nucleolus. All terminal nucleoli fused. The three chromosomes are each composed of two relationally coiled chromatids, and each chromatid has many gyres of tight major coils in it. Chromosomes are anchored to the centriole of the 5th flagellar band by the remains of the achromatic figure (chromosomal fibres).

FIG. 82. This is the same as the previous illustration except that two of the chromosomes are short (light) and one is long (dark). Both types of aneuploids occur in approximately equal numbers.

FIG. 83. Prophase of a 3-chromosome variety of *H. tusitala* which, by division, will produce two nuclei like the one illustrated by figure 81. It has four long and two short chromatids, each chromatid being composed of two relationally coiled half chromatids, and each half chromatid has many gyres of tight major coils.

FIG. 84. Two-chromosome or haploid variety of *H. diversa*. Prophase after resting stage. Achromatic figure has begun to grow following resting stage but has not made contact with nucleus. Four chromatids are present and each is composed of two relationally coiled half chromatids. (In *H. diversa* at this stage there are usually fewer gyres of relational coiling of half chromatids than in *H. tusitala*. Centromeres were not duplicated when chromatids duplicated and became half chromatids; hence each half chromatid has the same centromere. From the standpoint of centromeres this is a haploid; from the standpoint of chromosomes it, like all double chromatids, could be termed a diploid.

FIG. 85. Four-chromosome or diploid variety of *H. diversa*. Prophase just after resting stage. Achromatic figure lies farther away than in previous illustration and chromatids will appear to shorten considerably before it makes contact with them. Eight chromatids are present, each being made up of two relationally coiled half chromatids. The chromatids also have some relational coiling left in them. Since this is an earlier stage than the haploid shown in the previous illustration, the major coils of each half chromatid should be smaller, but I doubt that all the difference in size can be explained on this basis.

FIG. 86. *Spirotrichonympha mirabilis* from *Porotermes adamsoni* Froggatt from A.C.T., Canberra, Australia. Four prophase chromatids, each chromatid being composed of two relationally coiled half chromatids. Centromeres are median or nearly so. Super coils only shown in the two chromatids on left; in upper right chromatid, relational coiling of half chromatids is shown in both arms; in lower right chromatid, major coils are shown in both arms. It should be noted that when double chromatids develop super coils, the two half chromatids are incorporated into common super coils; they are not separate or individual as in the majors and minors.

FIG. 86e. Portion of the chromatids of 86 to show that minors are incorporated in majors, just as in *Holomastigotoides*. × 7,200.

PLATE 33

TELOPHASES (81,82)

PROPHASES
(83,84,85,86)

81

82

83

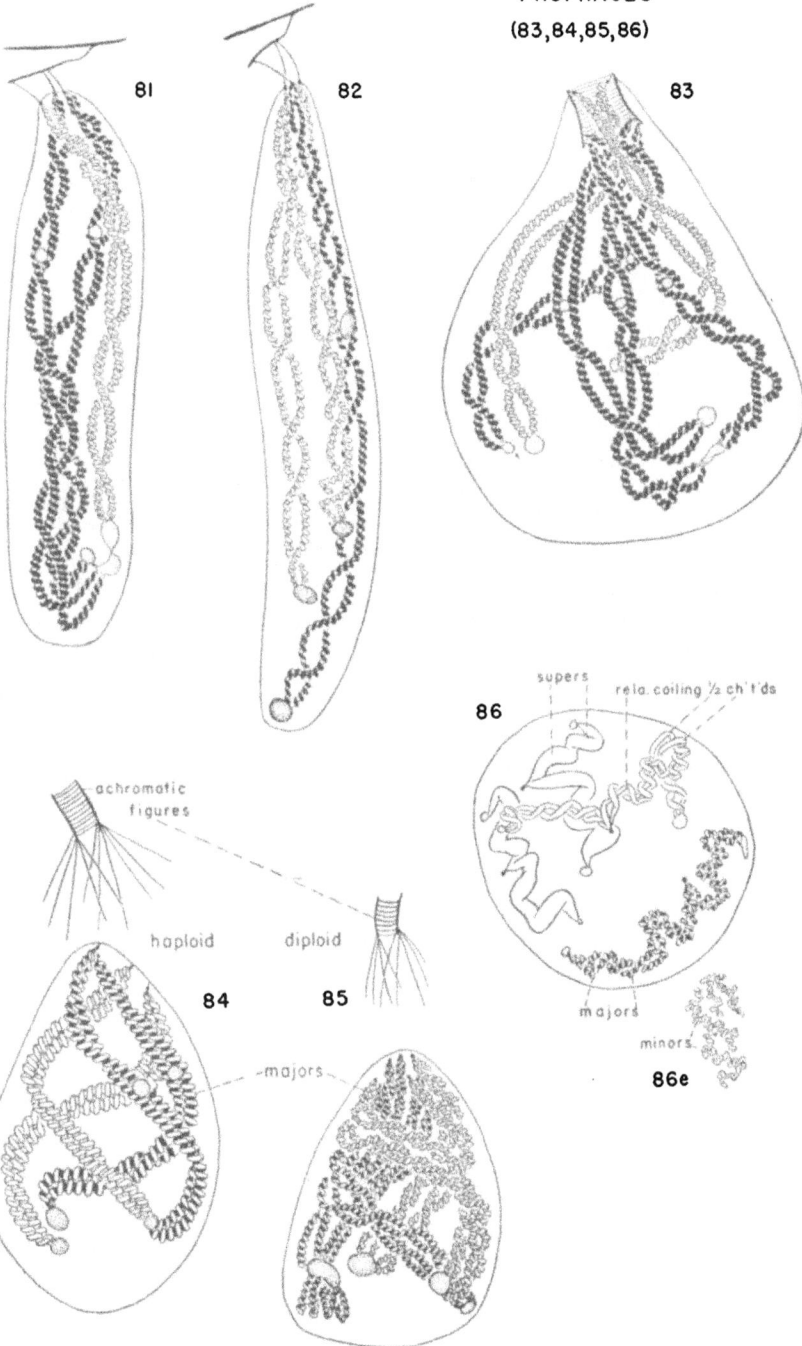

supers

rela. coiling ½ ch't ds

86

achromatic
figures

haploid

diploid

84

85

majors

majors

minors

86e

PLATE 34

These illustrations show how prolonged fusion of all terminal nucleoli sometimes retard the unwinding of relational coiling of chromatids and thus produce contortions.

FIG. 87b. All nucleoli fused. Both short and long pairs of sister chromatids distorted.

FIG. 87c. Since not all of the relational coiling of the half chromatids comes out in this generation, it has been distorted very little.

FIG. 88b. All terminal nucleoli were fused until a short time before this nucleus was fixed. One nucleolus and its chromatid pulled away. When freed, this chromatid rotated and thus removed most of the relational coiling between it and its sister. The nucleoli of the other pair of chromatids have remained fused and this pair (dark ones) not being able to rotate, has kept much more relational coiling than is normal at this stage in its life cycle.

FIG. 88c. There is no difference in the relational coiling of the half chromatids in either pair of sister chromatids. Hence the inability of the short chromatids to unwind did not retard the unwinding of the half chromatids; they have even more gyres than the half chromatids of the long pair.

FIG. 89b. Here the fused nucleoli have come apart so recently that some of the retardation of the unwinding of the chromatids may still be seen.

FIG. 89c. The unwinding of the relational coiling of half chromatids has not been retarded.

PLATE 34

rela. coiling of
chromatids

rela. coiling of
½ chromatids

87b

87c

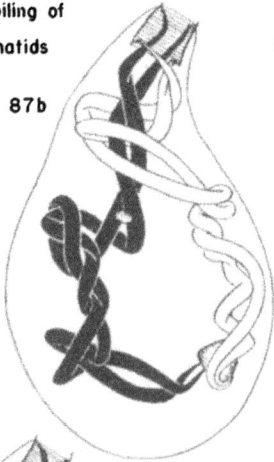
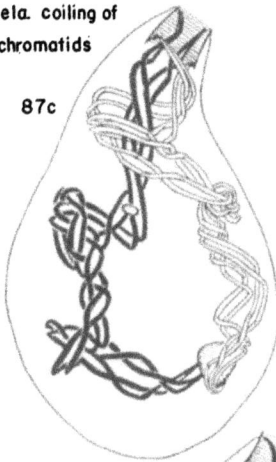

88b

88c

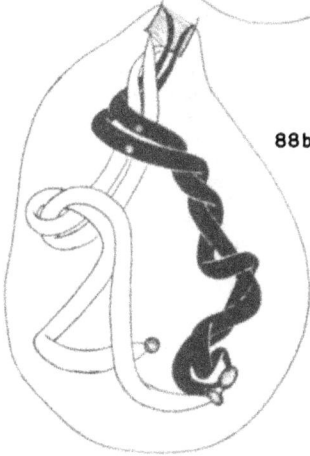
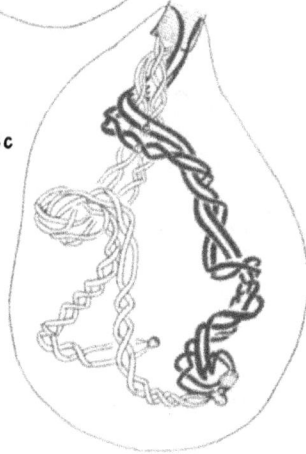

89b

89c

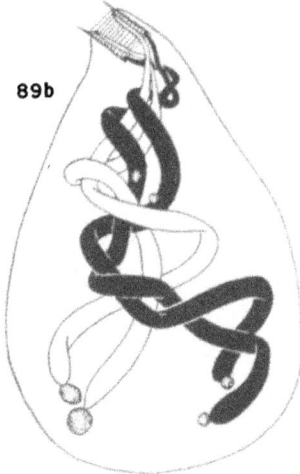
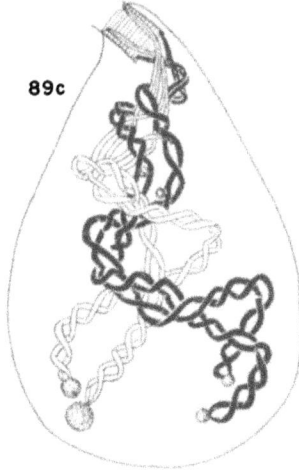

PLATE 35

FIG. 90. Telophase of *H. diversa* showing how in this longitudinal dividing species four flagellar bands unwind in one direction and four in the other, each set of bands carrying a nucleus with it. One set is viewed vertically; the other laterally. In this species individual membranes form around each chromatid in metaphase, the other material is discarded, and by telophase the membranes are fusing to form a new nuclear membrane. Centrioles are shorter than those of *H. tusitala*. Central spindle is pulling apart. × 1,500.

FIG. 91. Two posterior daughter cells attempting to fuse. Each still has a single flagellar band. × 740.

FIG. 92. An example of transverse division of *Holomastigotoides* sp. with two flagellar bands (from *Prorhinotermes*

simplex). One band, the upper one, is doing all the unwinding. When unwinding is completed, transverse division of the cytoplasm occurs and produces a posterior and anterior daughter of the same size, each daughter having a single flagellar band. × 740.

FIG. 93. A posterior daughter attempting to fuse with a late prophase cell with five flagellar bands. Sometimes three or four, and in a few instances as many as five or six, posterior daughters attempt to fuse with (enter) a cell like this one. The single and double-chromatid varieties always attempt to fuse with themselves, and neither ever attempts to fuse with *H. diversa*. × 740.

PLATE 35

90

91

92

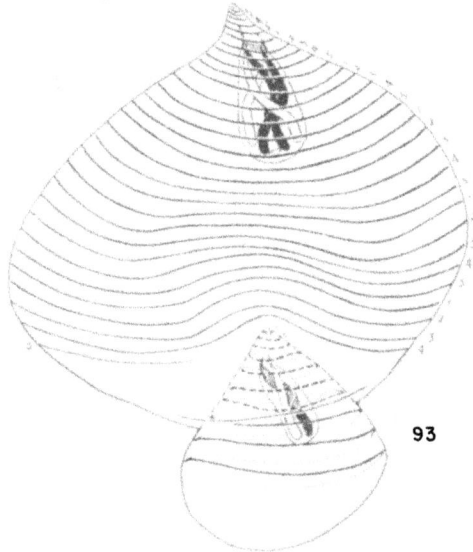

93

97

INDEX

Aberration, relation of "free ends" to nucleoli, 11
Achromatic figure, breakdown of, 10, 11
  connection to nucleus, 4, 8
  cycle of (in *H. diversa*), 10, 11
  description of, 8–11
  growth of, 9, 10, 15
  function of, 4, 8, 9 •
  inhibition of (artificial), 21
  movement of chromosomes by, 9, 15, 16
  occurrence of, 4
  origin of, 9, 16
  persistence of (in *H. tusitala*), 1, 4, 7
  resting stage of, 10, 15
  staining of, 2, 11
  unipolar, failure to function, 17
Anaphase (poleward) movement, failure of, in unipolar achromatic figures, 17
  relation to major and minor coils, 15
Aneuploidy, occurrence of, 4, 5, 6, 8
Astral rays, failure to join centromeres, in unipolar achromatic figures, 17
  function of, 8–12
  origin of, 9
  relation of, to chromosomal fibres, 9
  staining of, 2, 11
Axostyles, origin of, 7
  position of, 7, 8
  relation to nucleus, 4
  staining of, 2

Barbulanympha, centrioles, artificial removal of, 21
  centromeres, position of, 20

Central spindle, bending of, 10
  disappearance of, 10, 11
  function of, 8
  growth of, 9, 10
  origin of, 9, 10
Centriole, description of, 8–11
  development of, 4, 16
  effect of molting hormone on, 21
  in *H. diversa*, 8
  in *H. tusitala*, 1, 7
  origin of, 8, 9
  persistence of, 1, 7
  position of, in prophase, 4, 10
  relation of: to achromatic figure, 9, 16, 21; to centrosome, 8; to flagellar bands, 3, 7, 8
  removal of (artificial), 21
Centromeres, behavior of, 17
  connection of: to achromatic figure (in *H. diversa*), 8, 9; to chromonema, 20; to nuclear membrane, 11, 12
  continuous functioning of (in *H. tusitala*), 1, 7
  cycle of: in double chromatid forms, 17; in meiosis, 17; mitosis, 17
  "diffuse" or "polycentric," occurrence, 20
  duplication of, 9, 17
  duplication and function of, in meiosis and mitosis, 17
  function of, 25

individuality of, 20
movement of, by achromatic figure, 9, 16, 17
position of, 20
relation of: to achromatic figure, 20; to chromosomal fibre, 20; to chromosome, 20
separation of (before contact with astral ray), 16, 17
staining of, 20
terminal, 2, 6, 8, 19, 20, 24
use of phase contrast in studying, 12, 20
Centrosome, occurrence of, 8
  relation to centriole, 8
Chromatids, definition and discussion of, 2
  double, 5
  duplication of, 14, 18
  free (in cytoplasm), 12
  half, see Half chromatid
  in meiosis, 2
  in mitosis, 2
  matrix of, development, 18
  separation of, 8, 15
  single, 5
  types of, 1
  unwinding of, 9, 15
Chromonema, attachment of nucleolus to, 11
  bending of, 12, 14
  coiling of, 12, 13
  connection of, to centromere, 20
  duplication of, in meiosis and mitosis, 20
  elongation (apparent) of, 12, 14
  loss of "twists and kinks" of, 22, 24, 25
  relation of, to matrix, 12, 13, 23
  rotation of, 9, 14, 15
  staining of, 12, 20
  use of phase contrast microscope in studying, 12, 20
Chromosome, aberration, 11
  centromere on, position of, 2, 6, see also Centromere
  chemistry of, 19
  coiling of, 1, 3, 6, 8, 22, see also Coiling
  components of, 12
  connection to achromatic figure, 4, 8–12, 15–17
  cycle of, in meiosis and mitosis, 17
  "daughter"-, see Chromatid
  duplication, see Chromosomal duplication
  evolution of, 5
  fibre, see Chromosomal fibre
  identification of, 11
  major coils of, see Major coils
  matrix of, see Matrix
  membrane of, see Chromosomal membrane
  minor coils of, see Minor coils
  morphology of, 19
  movement of, see Chromosomal movement
  numbers of, in different stages, 2, 5
  of flagellates, comparison to those of higher organisms, 1, 19, 20

relation of: to achromatic figure, see Achromatic figure; to centromere, 20, see also Centromere; to nuclear membrane, 11, see also Nuclear membrane
removal (artificial), 21
resting stage of, see Resting stage
rotation of, see Rotation
shape of, 2, 4
"splitting" of, 16
staining of, 2, 11, 12
terminology of, 2
types of, 1, 5, 12, 16, 24
use of phase contrast microscope in studying, 12, 20
Chromosomal duplication, definition and discussion of, 16
  in mitosis and meiosis, 17
  occurrence of, 3, 9, 12, 14, 18, 22, 23
  relation of, to nuclear membrane, 12
Chromosomal fibre, definition of, 9
  origin and function of, 8–10, 16
  relation of, to nuclear membrane and centromere, 20
Chromosomal membrane, formation of, 11, 12
  relation of, to new nuclear membrane, 12
Chromosomal movement, by achromatic figure, 9
  during anaphase, 9
  during prophase, 9
  during telophase, 9
Chromosomal twisting, cycle of, 24, 25
  occurrence of, 14
  origin of, 18, 19
  types of, 13, 14
Coiling, see Individual coiling, Major coils, Minor coils, Relational coiling, "Relic coiling," "Standard coil," Super coils
  in resting stage, 19
Coptotermes, *Holomastigotoides* in, 3, 4, 5
Crossing over, in mitosis, 2
  relation of, to formation of nuclear membrane, 12
Cryptocercus, meiosis and sex of protozoa in, 18
  molting hormone in, effect on oxygenation, 21
  protozoa in, related to *Holomastigotoides*, 5
Cytoplasm, division of: longitudinal, 4, species with 4, 5; relation to nuclear membrane, 12; transverse, 4, species with, 5, 8; types, 4; unequal, 5, 8
  incompatibility of, in types of *Holomastigotoides*, 5

Diploids, functioning of centromere in, 17
  occurrence of, in *Holomastigotoides*, 5, 11
  size of, in *H. diversa*, 8
Diploidy, as regards centromeres and chromatids in *H. diversa*, 18

www.ingramcontent.com/pod-product-compliance
Lightning Source LLC
Chambersburg PA
CBHW081337190326
41458CB00018B/6027